T0157431

Order this book online at www.trafford.com
or email orders@trafford.com

Most Trafford titles are also available at major online book retailers.

Printed in the United States of America.

ISBN: 978-1-4669-6744-1 (sc)
ISBN: 978-1-4669-6681-9 (e)

Trafford rev. 11/09/2012

www.trafford.com

North America & international
toll-free: 1 888 232 4444 (USA & Canada)
phone: 250 383 6864 • fax: 812 355 4082

CAN DO IN YOTTABYTES

Charles O. Maul

CHAPTER 1

Technology emerged from the University Research Center
to Gladstone Electronics.
New energy efficiency, replaced old CMOS Technology
at their lab, in Crystalville.
Computers analyze global warming problems and human
disease. A small-computer
Molecular, Manufacturing Machine needs a large memory
unit called a yottabyte.

Gladstone dispelled fears, regarding the dangers of self-replicating nano particles, to help the cause of Global Nanotechnology. These findings resulted in the prototype, in a research lab at Gladstone. I took LH-1 Highway from Austin as soon as I got the call from CEO, Harold Remkin at Gladstone. The call from Harold Remkin, had nothing to do with my ability to trace an electronic circuit. I could not do that, but I could trace the activities of the young genius who had built the smallest super-laptop ever produced. He was missing along with his special laptop. The prototype represented the first step in building the kind of small computers necessary for molecular manufacturing.

The key-word is Molecular, as the scientists and electrical engineers all know. They are concerned with using the atoms and molecules as building blocks. This is called a Molecular Technology and it will change the world in many ways. It could bring about an industrial revolution. Some think it's only a few years away. Ray Kurzweil, has said in his book "Transcend: Nine Steps," that at least by the year 2034 we will be able to copy anything with nanotechnology.

My trip from Austin up the LH-1 took me right to my destination, in Crystalville.

It was a welcome change to be driving along in my own lane without cars cutting me off. I gained a relaxed composure that allowed me to enjoy the scenery, as I passed through heavily wooded areas, interlaced with developed ranch territory. And finally there was the view of Lake Crystal. I slowed down to read the sign for Gladstone Electronics, and turned off on the frontage road, which put me directly into the parking area. I continued on to the visitor's parking and parked in the first available space. I walked back about fifty yards to the polished, granite three-story building, where I had spotted the abundance of tinted plate-glass windows and large metal entrance doors. The first door I came to at the left corner of the structure had a sign attached that read—All Visitors Check-In.

"I'm here to check in." I said to the officious young woman sitting behind a console.

She adjusted her bifocals and smiled. "Who are you here to see?" She asked.

"I was told to go to the head of security."

She took my name and pushed a button on the console. In a few moments a bald,

Hawk nosed, man appeared in a blue-serge suit. He asked for my ID, which consisted of my Texas driver's license and my wallet-version of the PI license.

He talked to me for a while, watched my eyes when I answered his questions, and said for me to follow him. We took the elevator up to the third floor, got out, and went across the hall to a substantial oaken door with two permanent placards. They were one above the other. The one on top said C.E.O. Harold Remkin and the other was marked Dr. John Von Mueller. My well dressed security man opened the entrance door and led me into a wide corridor with side rooms to the right. We were met inside by a more glamorous version of the woman on the first floor, without glasses. She was a blonde, wearing a similar matching skirt and jacket with a spotless white blouse. She had a radiant smile that persisted even after it was turned off, and one of those fashionable dark tans. It was her confident smile that won me over. She flashed it at me as she reached for my hand. "Marsha Brooks here," she said. "And you must be Dan Brandon. Your guide was Carl Watson, in case he forgot to say."

"He did forget."

"The man is cold, or he thinks part of his job is to intimidate people."

I nodded in agreement as I filled a cup with hot coffee and looked back at her.

"Just be seated there at the conference table while I get Harold and Doctor Von."

There was ice water in a spigot next to the coffee pot and I put some of it in mine.

Waiting till Marsha returned, I sampled the coffee, and got a paper napkin to fill it with enough food to make my lunch, in case things happened at such a rapid rate that time became a premium commodity. It was about nine A.M. but I only had coffee and toast for breakfast. I filled the napkin twice, tossed out my garbage and was standing empty handed. Marsha returned with a man on each arm. The older graying man on the right advanced ahead of the others with his outstretched hand reaching for mine and we shook. He wore an expensive suede-jacket, light-blue shirt, with a gray and white striped tie.

"I'm Harold Remkin, we talked on the phone," he said with a flourish. "I'd like you to sit in on this Marsha." As we were being seated at the conference table he glanced around at us to assure us that he was in charge and would be for a long time to come. The other man in a black-serge suit with a white shirt and blue tie closed the door and locked the dead bolt from the inside. He came toward me with his hand out. "John Von Mueller," he said as we shook. "Just call me Von like everyone else." He motioned for me to return to my seat next to Marsha. She had seated herself on the left side of the table between me and Remkin. I was on the corner and they had reserved the spot at the head of the table for Von Mueller. He joined us with his back to the door.

"Harold wants me to go over some of the technical aspects of our situation," he said.

"I'll also be filling you in regarding the incident Harold mentioned during his cell call. I'm sure you understand we're trying to keep it quiet. We put out the word that Bob Clauson is on a personal leave." Von folded his meaty hands on the top of the table and looked down at the rest of us. Well over six feet tall and a good two hundred and twenty pounds he towered over us even in the sitting position. He smiled and glanced over at Harold. His hair was dark-brown, not frosted like his boss. Von was handsome.

The pause was long enough to allow us to adjust to the power switch. "We are the only ones who know he's missing," Harold Remkin admonished, "and that our laptop, prototype is gone. This morning, when we started looking for him, I found the laptop was missing too. That was when I called you, Mr. Brandon. Since I've heard of jobs you've done for other high-tech companies in our area, I felt you would be discrete."

I had never had a problem about my discretion, but there were a lot of other things that had to be worked out before I could go into action in behalf of Gladstone Electronics. "The police were also involved on other cases," I said. "This doesn't sound like the sort of case where an investigator should be snooping around alone."

"We're prepared to offer you an exclusive arrangement and you can bring in your own operatives, as many as you need. They'd be sworn to silence as you would be."

"Even with plenty of my own operatives I don't think I could get the kind or results you're thinking about. For something like this you need the FBI and probably the CIA."

"We've thought of all that," Remkin said. "We don't need to let our competitors know we're at a disadvantage now. Can we try it first with your agency? We could get lucky with you calling the shots and not need to call in all those troops. We can't take the negative publicity that would come out of it."

"I appreciate your situation, but I think you're expecting too much from me."

The power switched again and I got the nod from Dr. Von and the all knowing smile. Harold squirmed in his chair and came up with his own smile. It was back in Von Mueller's court. "We trust you that much, so why don't you trust yourself," Von said.

That Dr. Von was tricky. He knew how to put the bee right on your back.

"I'll try it for a day or two, but after that I may walk away, no kind of contract to worry about." This time Von Mueller squirmed in his chair and it was back to Harold.

"Consider it agreed to," Harold said, "and Von can get back to technical aspects."

Von was larger, sat up straighter, and had a PhD. I thought he was the true leader. "You mentioned the FBI and I was reminded of an FBI case involving a nuclear submarine," Von said. He paused until he had our entire attention. "Information regarding national security about the sub had been typed into a laptop and the laptop came up missing. They put up a $25,000 reward and they got their laptop back."

"Are you thinking money can be thrown at me in the same way, and I'll get back the laptop for you?"

"Yes, we hope so," Von Mueller answered. "It's a similar incident to our own, because our prototype laptop contains such important information that we're unable to go into production of the new module without it. You see, the new module involves nanotechnology. That's why it's a super-laptop. We've been engaged in the research for many years and it's now come to fruition. New computers today might have a terabyte of information, but our laptop has a lot more. It was put together using advanced micro-electro-mechanical-systems or what we refer to as MEMS. We started the innovation with quantum computers, and small super-fast electronic components. We experimented with nano-scale compounds and developed a safe module that has surpassed even modern tetrabyte technology. We're into, what we call yottabytes. We call the units of information yottabytes, but the memory space is about one nanobyte. We don't call them nanobytes but the memory contained inside our laptop is comparable to what a nanobyte of information would be. It's about a billion bytes."

I was still of the opinion that they needed more than a dolt like me who was color blind and couldn't even trace the most simple electronic circuit. How was I supposed to understand quantum computers? I didn't want to show myself up so what I said was, "All you really need is the prototype, not the inventor."

"That is not exactly true, Mr. Brandon," Von Mueller said. "We also need Bob Clauson. I worked with him on the prototype and was considered in charge of the research, but I'm not a completely devoted research scientist like Bob. I don't know enough about the prototype to go into production without him. He has no doctorate, just that little edge of expertise to make it work."

Well so much for my theory that good old Bob the goofy genius had suffered an attack of amnesia, or was still sleeping off a wild drunk. I had hoped we could do without him. I was going to be looking for Dr. Bob who was only Bob not Dr. Bob, so he could show Von Mueller how to produce the special computer. Bob was the man and the others were just the suits. They were off to cocktail parties while old Bob was at the shop doing the work.

"I bet Bob slept in his suit, if he owned one, or came to work in bib-overalls," I said.

They all smiled looking at each other, and Marsha even tittered a little out loud. Harold Remkin reached under the table and removed a few things from a leather case. He handed me a five by seven inch color photograph of a man who wore jeans and a matching denim jacket. Yep, that was Bob Clauson. Remkin also gave me a copy from Bob's folder at Gladstone. I had all of their business cards with cell numbers, when Von Mueller unlocked the door. He and Remkin were out the door in a second, while Marsha Brooks hung around a while to let me know she was single and ready to tango. She informed me that she had just reached forty and had already had a divorce. I did find her warm and attractive, so I had to be nice. But my girl friend wouldn't like it and I couldn't afford to get involved with someone in Marsha's position. I would need to work closely with her, so I just strung her along. I got another cup of coffee and we sat down and talked about things I needed to know, like what about Bob. We looked over the copy of his file I had been given.

"You cracked me up when you asked if he came to work in bib-over-alls," she said. "You were right on, of course, since they needed him so badly they had to let him do whatever he wanted."

"It says Bob is only twenty five and single. He owns a condo and drives a rebuilt 1955 Chevy, converted for the use of a non-petrol fuel, with a light nanotech body and electric engine. Were his parents contacted in San Antonio? I asked.

"Yes. His mother said he hadn't been there to visit for about a year."

"It doesn't say anything about favorite hang outs here in the file. Where do I start?"

"He talked about places around the drag," she said.

CHAPTER 2

Driving to Bob's condo, up the driveway, ignoring the vintage Chevy, I walked around the perimeter of his lot. That was what cops called "the walk around." You assume there are some people inside the buildings who may see you through the windows and come outside to talk to you. At least, that was the idea. Back at square one I went up to the restored vehicle like it was the first one I had ever seen, kicked the tires, ran my fingers over the olive green paint and tried the doors. I found the door unlocked, opened it, and left it hanging open for dramatic effect while I stepped inside for a look. It didn't look like one a car thief would want to steal, because it was clean, neat and orderly. That kind of car stands out from the average cluttered up mess like my own. I raked my hand under the front seat and came up with a few scraps of paper and a book of matches. The cover of the matches said El Dorado with an address on Guadalupe Street, maybe on the drag. I got down and used my pocket flash light to be sure I got everything. There were three crumpled receipts from other ends of town but I kept them anyway and put them in my pocket with the matches.

Shagging around to the over-head door I grabbed the handle and pulled up on it, but it didn't budge. I had the urge to quit while I was ahead at this location and race over to the drag, but this was not the right hour to visit the El Dorado. I tried the front door and it was also locked. Pounding on it as loud as I could like a bill collector got some results. A casually dressed young man came out the front door of his condominium. He carefully stepped down a few concrete steps leading to his rural mail box. I thought he was going to check the mail but he kept on walking toward me. "This dead beat been around here?" I asked.

"No, I guess not." he said. "Why do you call him a dead beat?"

"Oh, you know him then" I bantered.

"Yes, he's my neighbor" he said. "I don't like to hear noisy strangers come here and bad mouth him. Some of us work night shifts and sleep in the daytime."

"Does that explain what you're doing here at noontime?"

The stocky, blond man shrugged his shoulders and walked back up the stairs. "I'm calling in a complaint on you, so if you're still here the police will be checking you out."

When he went back inside and closed the door I got in my car, rolled the window down and got comfortable. I sat there for about a half an hour thinking about this case. What little I had learned was going around in my head. I could see when I had looked through the window of the over-head door that the inside of Bob's garage was neat and spacious. There was no car parked in the garage, so I reasoned that Bob could have acted in several different ways before he vanished from the earth. He might not ever bother to park the car in the garage, so he would have parked it last night in the present position and gone inside to sleep for the night. He got up this morning and locked the house, but he did not use the car. He either went somewhere on foot or someone else came by in a car, gave him a ride or abducted him. He might not have parked the car there last night. He could have met foul play and been abducted anytime after he was logged out Friday afternoon.

Whoever was behind it could have driven Bob's car to that spot and left it. Then a third party drove another car there and picked up the accomplice. I believed there were at least two involved in his disappearance. If Bob was involved in the conspiracy he would have needed someone to drive his car back and park it in front so it would look like he had been kidnapped. I had to change my focus. I drove across the road, around the short spiral driveway and parked behind the good neighbor's car. I was all smiles and chuckles as I knocked on his door. I wanted to be relaxed and ready to respond if he attacked me.

The door popped open and I took a step back. "I'm sorry I pushed you so hard, but I'm in an emergency situation. Your neighbor, Bob Clauson, really does not owe me any money. I'm an investigator working with security at Gladstone. They hired me to check out the neighborhood because something is not right around here. I can't tell you what it is but you seem to be around at a time when you might have seen something that could be helpful to us."

"You mean you're here to question me, not Bob," he said. "Yes, and when you said you were going to call the police I had to wait and see if you did. That would have been what I liked. You can still call them if you want to."

"No, I was just trying to get you out of my hair. Can you show me some I.D.?"

I flipped out my wallet sized version of my license. "You have to cooperate with me or I'll give you to the police. You can start out by showing me your own I.D. please." When he handed me his driver's license, I jotted down the number in my pocket notebook. I also wrote the correct spelling of his name. It was Marvin Shillar. I also got his phone number and work information. He was a musician who played the piano in local nightclubs. It explained what he was doing home at noon. He left me standing outside and I could hardly blame him considering the kind of treatment he had received from me. "Well Marvin, tell me what has been going on around here lately." He paled.

"I was wakened early this morning when three men started banging around over at Bob's place. They drove a white vehicle with a blinker on top, so I guess it was some sort of security outfit. One of the men was a locksmith. He opened the door. They gave me the same kind of story you just gave me. The bald headed one got in my face and I thought he would hit me, when I said the last time I had seen Bob over there was Saturday afternoon. Bob was washing the car in his driveway and I waved to him as I was driving out to visit a friend."

"You said they woke you banging around and then went inside and it was quiet enough for you to go back to sleep for an hour. You must have looked at the time to know it was an hour's sleep you got."

"I looked at the clock when they woke me the first time and it was seven o'clock. It only took a few minutes for the locksmith to get the door open. Then I went back to sleep, and slept until they were at my door at about eight."

"How well did you know Bob Clauson?"

"Pretty well, we have both lived here about three years and have tilted a few brews together."

My opinion of Marvin was raised about two hundred per cent, because I had finally found someone who would qualify as one of Bob's friends. I might be able to work this case and make a few bucks. I questioned Marvin a little further and he told me that when Bob started out at Gladstone he was able to set up his own work hours. Those special genius types were encouraged to exercise their creative abilities and come in to work at any time they felt productive. Bob was one of those people who woke up about four thirty or five every morning, got his cup of coffee with toast or a donut and was ready to work. He was at Gladstone between five and five thirty to check in with the security guard about every morning. They did have a clock there but it was for the hours of security provided for the laptop, not for Bob. They just wanted the security guards to keep track of when he came in to work and know where he was so they could protect him if they needed to. Security had really screwed up big time. I was certain that the prototype laptop would have remained in the lab from whatever time they had decided it was more than state-of-the-arts technology. I wondered how someone could have got it out of the lab past the guards.

"Did Bob ever tell you anything about the special research project he was working on?" I asked.

"No, he never talked about it, but I knew they had trade secrets over there,"

"Did you ever go with him to a bar for beers?"

"No."

"Did you meet any of his friends at his place?"

"No, we would just invite the other one over like that occasionally, when we'd run into each other outside. It was always late afternoon, when I'd be getting up to go to work my gig and he'd be home from his job."

I thought to ask him where he worked his gig and he said at El Dorado. That match-book in my pocket seemed hot enough to catch fire and burn my pants up, so I took it out of my pocket and toyed with it while I spoke. "Bob ever pay you a visit over there?"

"He wasn't the type to hang out in night clubs, but he did come in once with a dinner date. He introduced her to me, a real knock out, from where he worked. I think it was Martha or Marta . . ."

"Could it have been Marsha Brooks?" I interrupted. She had pretended not to be on any sort of intimate terms with Bob and it was hard for me to believe it. Why would she deceive me?

"Yes that was the name. She was dressed in one of those matching skirts and jackets, but Bob was just dressed casual like always. It was Saturday night and I would have even considered renting a tuxedo for a woman like that. She was wearing a corsage."

"How long ago was that?" I asked.

"It was just this last Saturday."

"You said you hadn't met any of his friends."

"Just have to be sure," I said. "Not cross questioning" I could do that later. It seemed that he had learned to be wary of cross questions.

"It's been a hard day for me and I still have to go to work."

I decided to let the poor man get some sleep. "Go back to bed Marvin," I said.

I had some cell calls to make and I could listen to the birds and stay off the highway for a while. I leaned against the side of my car and tapped Marsha's number. When she answered I asked why she didn't tell me she had dated Bob.

Her tone sounded irate. "It was just that one time, because I felt sorry for him."

She said she would be more forthcoming. I asked her to inform Harold Remkin that I was calling in one of my associates to interview Bob's parents and friends in San Antonio, and I would go into the case under the same verbal terms and conditions we had agreed upon.

After doing some of my own special research and making a few phone calls to ask about Bob I learned that he had been associated with the Nanotechnology Research Institute at the University I called the Department head of the research division where Bob developed the super-laptop. And he gave me the directions I needed to get over there and the name of the man in charge was Dr. Wallace Norton. It's a very large University and it helped to feel at ease there, which I did. It was a good thing I had taken a few courses in night school when I was a rooky cop in Austin. In about twenty minutes I made the trip down Highway LH-1 and was in UT territory.

CHAPTER 3

There was a uniformed guard motioning for me to come on into this hole in the wall place. There was barely room to turn around and get out of there. I didn't think this was where I was supposed to go. He motioned for me to stop next to a late model car that was parked up against the building. There were no marked off parking spaces, but the other car had a university parking sticker on the windshield.

Only a few feet from where I parked at the back of the building a large man dressed in a gray-serge suit waved to me from a small cave-like alcove.

"Come this way Mr. Brandon," he said. "I'm Wallace Norton."

I left my car with the guard and followed Norton inside the concrete cave. We walked for about a half mile deeper into the lighted hole. He was over six feet tall, about the size of Dr. Von Mueller, and I was reminded of how this was the way a scientist looked. They didn't look like the frail mad scientist type that we read about when we were youngsters reading comic books. Those of high intelligence were actually larger and stronger than the average person. They tended to look more like a football player or a basketball player. Norton looked to be in his early fifties, or late forties. He had a high forehead, light-brown hair, broad nose, square jaw and he was definitely another one of the suits.

We had been walking down a lighted concrete ramp, descending with stone or concrete on all sides. I was getting a slight case of claustrophobia, when we came to more of the red blinking lights on yellow two by fours. There was an "off limits" sign as the shaft reached a dead end with a locked steel door. Norton produced a key, unlocked the door and motioned me inside to a large room with comfortable leather chairs and vending machines. On the far side of the room was another metal door with a sign on it that said authorized personnel only. He opened it, stuck his head inside and waved to someone. In a few minutes a young sandy haired woman came through the door. Norton introduced her to me as Nelda Whittington. Under the white lab coat she was wearing jeans, a white tee-shirt and white sneakers. Her hair was short and straight. Norton brought us canned colas and we sat in the leather chairs drinking them.

"I brought you here to talk to Nelda because she works with Bob Clauson," he said casually. Nelda smiled and I thought she was attractive in spite of herself.

"Mr. Remkin called here early and left a message for me to call him back," she said.

"I came in at eight A.M. and when I called him back he said Bob didn't show for work at Gladstone this morning."

"Sometimes he would come here first to get something he had prepared at this lab, to make some improvement on the product at Gladstone, you see Mr. Brandon," Norton added. "I'm sure that Remkin was hoping Bob was here."

"Is there something here Bob might have come after, Nelda?" I asked.

"No there isn't," she said, "and if he does come here first he always hangs around long enough to chat with me."

"Have you taken a lot of the same classes with him?"

"Yes, way back to undergrad math and science classes," she said. "We were like brother and sister, wish it were more but it wasn't."

"Did you notice anything different about the way he was acting lately?"

"No, and I'm the one that would have. He talked to me about the hang ups of his parents and everything else."

"Do you think he could have been stressed out and just taken off on his own?"

"He loved his work too much to do that." She said. "He's a compulsive worrier and I mean about the job, to the point that nothing is ever just right. He comes to this lab sometimes and tells me about how he woke up from bed and solved some problem we were working on here. I don't think he'd be able to run away from the job."

"I agree with Nelda," Norton chimed in. "He's always been loyal to Gladstone and I'm sure he would never sell out to another company."

It seemed that Nelda had good rapport with Bob. If anybody knew him she did, and she was included in the short list of people who were aware of the fact that something had happened to him. I tried the silent treatment for a while, looking from one to the other. I was determined not to speak until one of them broke down and spoke first.

"Something bad has happened to him, I'm sure of it," Nelda said with a shudder.

"The question is what?" I said. "I need a few suppositions."

"Well, he's the kind of guy who might rub somebody the wrong way," Norton said.

"Yes, he could fill up several black boards with formulas, but not good at social skills," Nelda said. "After one class some guy tried to hit him and Bob had to block it."

"So he could have been a victim of random violence I hear you saying."

"He could have been kidnapped to get technical secrets from Gladstone," Nelda said. "And we agreed Bob would never voluntarily leave Gladstone to sell us out, so that would mean that he had met foul play."

The summary of all of the suppositions seemed to be either some kind of random violence, or a conspiracy which was aimed at separating Bob from his beloved company to gain technical knowledge. It had been worth the trip down here to learn that Bob had a bad habit of saying the wrong thing at the wrong time. When I thought about this I came up with a new supposition. What if it was a combination of both things and someone who didn't like Bob was used as a cat's paw. That person might not have even known about the conspiracy and had just wanted to take him somewhere and beat him up. When I had questioned them as far as I could, we exchanged business cards with cell phone numbers. I asked them to work up a list of enemies like the one who took a swing at Bob. Nelda thought one of her classmates might be able to find out where that guy was. They have ways of running down people at the University to see if they're still taking classes or if they quit UT to transfer to another University or just to take a job.

"Well, you know where to reach me from eight till four," Nelda said. "I carry the cell phone all the time. It's my only phone at my apartment, and I'm available twenty four seven to do anything that might help. As I said He's like my brother."

I said I appreciated it and would call if I thought of anything she could do. Norton squinted behind his rim-less glasses and said "take care." That was all he said, and I guessed that time segments were always filled for the PhD suits. He unlocked the outer door for me and I went back through the tunnel to my car.

I got a tuna salad sandwich from one of the vending machines before I left, and took it with me to the car. I ate the sandwich in front of the uniformed guard, because I never eat or talk on a cell while driving. I had the window down to air out the car and the man came over with a smile. "Take your time, it beats having a wreck," he said. I told him my other rule. There were times the phone rang while I was driving and I answered it. That was the only time I broke my own rule and I usually pulled over to talk, if it was a long call.

By the time I pulled onto LH-1 it was dusk and I was glad I had that cola to get caffeine. This time of day between the dog and wolf I usually felt more like the dog and a tired one. I often extended myself on coffee, when I was in a situation like this. If someone's life might be at stake, my conscience forced me to keep going till I dropped. I didn't save any lives if I went to sleep at the wheel and lost my own.

The highway lights were on now to remind me to switch mine on. I was passing scenic ranch territory, but it was only shadows at night. In a few moments the three-story Gladstone building loomed out above the shadows on a well lighted hilltop.

Not until I was safely parked in the lot next to the building did I touch the cell, but I needed to touch down with Harold Remkin before I started on Carl Watson. I pecked out the number and waited. When he answered I asked him if he got the message from Marsha.

"I was glad to hear you were coming with us," he said.

"There's one more condition I need met, Mr. Remkin," I said. "Please, have Carl Watson to level with me, and disclose everything he's done on this case."

"Are you prepared to do the same with him?"

"Of course I am, but he has got to fill me in on what he's already done," I said. "Tell him I'm right outside the building and on my way to the first floor entrance."

"Yes, I'm calling him now on the fast line and he should be there in five minutes."

Hustling along the distance between my car and the building put me there a few minutes before Carl arrived all red faced and tight as a band saw blade.

"What took you so long, Carl," I bantered. He didn't seem to take it well.

"You got me crosswise with my boss, wise guy."

"Just do what he asked and the pain will be gone, till the next time you have to level with me."

"He said you're on the case now, but I'm saying get off it."

"I'm already involved in it, too much to walk away."

"We better go outside for some privacy then," he said hotly."

"It's private enough over there." I motioned to a corner where there were several chairs set apart from the rest of the room and I felt some of the heat leaving him. He could see we were wasting a lot of time and so he walked over there and sat down in one of the chairs. I did the same and in a few minutes we were relaxed enough to talk normally. He admitted that he had a reasonable relationship with the police and had managed to get the airport and roads out of town covered, as soon as he heard from the lab guard that Bob had failed to show up for work.

The loss of the super-laptop was a different matter. He didn't seem to have a clue, and I understood why he was so touchy. Carl was going to loose his job if Gladstone had to suffer the loss. I had to convince him that I would be a help, not a hindrance and I worked at that for a while until I could see he was starting to buy it.

"I have an operative checking with Bob's family and friends in San Antonio," I said. "Willie Harland has his own agency there and things are slow now. He said he could send a couple more up here if we need them."

"We need the kind that can search with nothing to go on and make things happen."

I assured him they would be that way, and I told him about what I had found out already. He seemed to like it, but he kept drifting back and forth between what we were talking about and his own insecurity.

"I suspect one of my own men of being in on the heist of the laptop," Carl lamented.

"You don't want a uniformed cop to watch him and it needs to be someone he doesn't know."

He went for the idea and I got Willie Harland on my cell right away. Willie said his wife, Helen, was good at interviewing people, so he had turned that job over to her. I had used her before and thought she was better at it than anyone else I had ever heard of. Willy was coming here himself and bringing Ray Dartmouth, who was also good at any type of surveillance. They would report to me as soon as they got here and the meeting place was on the drag at El Dorado, where I could check out the place while we got together with Carl. I had gone by the restaurant on my way to the University but didn't have time to stop there. It had been too early in the day anyway. It was late enough now to check out the evening clientele, so Carl agreed to take his own vehicle and meet me over there.

Back-tracking again on LH-l, I was driving on the south highway in a few minutes and got off at the Lamar Blvd exit. In ten more minutes I pulled into the short turnaround in front of a tan colored adobe building with an old time neon sign that said El Dorado hanging across the window. I parked at the curb and went inside, where I found Marvin Shillar sitting at a table next to his piano taking a short break. Two elderly couples seated at the next table were laughing and talking about some of the old songs Marvin must have been playing. I went to the bar, ordered a sprite cocktail, and another brew for Marvin. It was delivered to Marvin's table and I joined him. When he saw the replacement for his empty longneck he seemed to remember me and smiled.

"Been sleeping better lately, Marvin?" I asked.

"Yes, thanks, Mr. Brandon,"

"You remembered my name, and you were only half awake when we met."

"I think you have a way with getting people's attention, and thanks for the beer."

"The least I can do, since I may sit here most of the evening nursing this sprite along." I didn't tell him how I had called a meeting here and the other members would be coming along shortly. In fact, Carl Watson, just popped in the door, spotted me and came over to sit down next to me. He anxiously watched the front door and completely ignored Marvin.

"I believe you two have met," I said. Carl looked like I had made him ill and still said nothing. He looked at his watch and loosened his tie. I thought if Marvin went back to the piano. Carl might do something to him, maybe pull out a deputy's badge and arrest him. And thanks to divine providence, for small favors, Marvin got up and walked somewhere to the rear of the bar just as Willie and Ray walked in. He wouldn't be invading the place with music for a while and it would only take a few moments for me to make the necessary introductions. I didn't even do that, because Willie went directly over to Carl and introduced himself. Ray did likewise and Carl was ready to show them the residence of his suspected employee, and get them started on the surveillance. I let them go and do their thing, while I took a sip of sprite and considered my own immediate situation. While I was here I wanted to talk to the bartender who seemed to have a little slack time now.

She was also the restaurant manager, and was flitting back and forth all around the place, slender, attractive and fully in charge. I asked her who had the duty last Saturday evening and she said she did, with the help of a part time man who assisted her during the busy time. The beauty of having a woman in charge is that women have more options available when it comes to dealing with violence. The studies have revealed that they may step aside and shed a tear for a few moments, where a man if left to his own devices might just crack a skull.

"Did you notice a couple here for a dinner date, Saturday?" I asked.

She said she had and they left early.

CHAPTER 4

Nelda had given me Junior Gilbert's address on the drag, not far from El Dorado. It couldn't have been more than a few blocks farther on Guadalupe, so I walked. In the close proximity of any university there are always some people who are coming and going like students but are not students. They may even live there but do not belong. Some of them looked at me with guilty expressions. I thought they were looking for a chance to steal something. They thought I belonged.

It was a small, dirty-white frame house, cheap rent but not a frat house. I came to the south side of it first going north on Guadalupe and moved along the front, which faced west. There was a narrow alley on the north side and I turned on it and went to the back of the house where there were two vehicles parked in the back yard. One of them looked like it might run and the other one was up on blocks. It was dark inside, but I remembered the front door had been open so the breeze could blow through the screen door. I went back to the front door and rapped on the screen door. The name I had been given was Junior Gilbert and I called his name. There was no door-bell to ring and my voice sounded louder and more threatening than I would have liked. With a name of Junior I thought the guy might be dangerous. He had attacked Bob on one occasion and I had seen in Bob's office file that Bob was six three and two hundred and twenty five pounds.

When nobody came, I pushed the screen door and it came open. I cautiously stepped inside and it was easy to see the layout of the place and verify that no one was home. It consisted of living room, kitchen, bathroom, and two medium sized bedrooms. One bedroom was for a man and a woman with modest furniture and clothing. I was more interested in the other room and looked through it carefully.

I slammed the screen door on my way out so it closed tight as it had been before. As I returned to the sidewalk in front of the house, an older model half-ton pickup truck pulled up by me and a man and a woman got out carrying a few bags of groceries.

"Can I help you?" The man said as the woman continued walking through the yard and into the house.

"I'm looking for Junior Gilbert," I said. "I knew him from classes at U. T. and wondered what he's doing now"

"He's my younger brother, had to lay out this semester when he ran out of money. He lives with us and works to get money, so he can stay in school."

"I might be able to help him with that."

"He's looking for a job now."

I flipped through my pocket notebook, found an empty page, wrote Bob Clauson and added my phone number. I tore it out and handed it to the man.

"Have him give me a call at that number," I said.

He nodded and said he would, as he made his way back into the house. I walked back to my car in front of El Dorado and called the number for Tim. I got no answer, so I went back in the restaurant and found the manager. She was cleaning off the counter top of the bar. Marvin was back at the piano doing a modified version of "Autumn Leaves" He was not pounding the keys too hard, but I had to talk loud for her to hear me. She stopped busying herself around the counter and came over to me.

"I tried calling your friend, Tim, but he wasn't home," I said. "I thought you might give me a description of the young man who got into it with Bob."

"Well, he's been in here before, so I should be able to. I'd say he was about five, foot ten and a medium build, a little stocky, maybe two hundred pounds."

"What color for skin and hair?"

"He was a kicker with a good outdoor tan, and dark Brown, almost black hair. The hair was combed back long on top but with a normal trim on the sides. I'm not sure about the eyes, think they were brown and he had a big flat kind of nose."

After she got to thinking she remembered his features very well, and I told her so. The small crowd was starting to sing along with the next selection, which was another old familiar one. I went back to the car, drove back to the dirty white house, and parked in front. I sat there with a road map spread out in front of me, so I could seem to be consulting it if someone came along. In an hour a young man fitting the description I had just been given drove a used motorcycle down the alley and parked it in the back of the house. I was in an exposed position parked at the front curb under a bright overhead city light and I wasted no time getting out of there. Junior Gilbert had been into a little altercation with Bob on two separate occasions. The ringer of the cell phone was on. I was anxious to see how he talked, if he called back thinking he was calling Bob. If he had not been in on the scheme to abduct Bob he might call me to smooth things over with Bob to get a job. If he had been in on it or been used as a cat's paw he would call the number I had given his brother. I wondered if Junior Gilbert's brother had remembered what I looked like well enough to describe me to Junior. That was something for me to keep in mind.

In a few minutes I was on North Lamar heading south, on my way to Claire Prevost's downtown apartment. She had said she would be there all evening and I took her at her word. I kept an office at Southland Realty, not far from her anomalous apartment complex, for a number of years. Very little business is generated from it.

The parking space in front of Claire's ground floor apartment was vacant and I parked there and let myself in with the key. I thought the open space in front meant she was not home, but things never turn out the way you think they will. I was glad about that. She ran to me and we locked in an embrace as soon as I went through the door. I carried her into the bedroom and we culminated what we had begun, but with much less speed and more enjoyment. Her pleasant oval-face got glued to mine.

"Good thing you showed up," she said. "I was about to call an old friend."

"Oh, you were, Oh really."

"No, O'Reilly."

All the pent up stress of the case was gone, and we were laughing, as we had regressed to childhood. She produced a fresh bottle of sangria and filled two wine glasses. We sipped it and talked incessantly, about nothing, till sleepy and fell into bed.

I started my second day on this case with Claire's special breakfast tacos and a renewed sense of devotion to my new client, Gladstone Electronics. She was back to her own job as legal eagle for Bradley Dormer and White, an Austin law firm. In a few minutes I was back on Lamar Blvd., one of the old standby roads going anywhere north or south in the city. To access the central part of town the way I had been doing it was the best way to go. The traffic lights were timed for folks like me, who went right along at about forty miles per hour with few stops. The cross streets going east or west were a different matter. There were a lot of them to choose from, and you had to be selective or risk the chance of a major screw up. I took MLK Blvd. to get past the U.T. bottleneck and to northwest Austin. Then I got back on L H-1 and took it to Crystalville.

When I got off the freeway at the Gladstone exit and parked, I checked the cell phone and found it was still on. Junior Gilbert had not called, so he must not have wanted a job bad enough to get one with Bob's help. It could be that he was on to me now. He might know exactly where Bob was, and that he would not be cruising around on the drag, leaving messages with big brother. I put the cell phone back in my pocket and took the elevator up to three. I got off alone, rapped on the door the way Watson had done and then opened it and went inside.

There was a small corner office I hadn't even noticed before. Marsha was sitting there behind a computer watching whoever might go by. She seemed surprised to see me and I had to do something about that. I nodded and crossed the long room to join her in the small office.

"Are you an MBA Marsha?" I asked respectfully as I dropped into a computer chair.

"Yes, I finished it last year and landed this job because they needed one here.

"I think something happened to Bob after your dinner date Saturday. Did you know any of the other people at the restaurant that night?"

"Yes, my ex husband was leaving as we were going in and he made an unfortunate comment to me. After that I wasn't very good company for Bob. It was something to the effect that I was going out with very young men lately."

"Marsha, I understand," I said. "That was a stupid thing for him to say and not true. I'm sorry I had to remind you of it."

"It's OK. You were just doing your job."

I still felt like a cad. At some other time, after we were over the embarrassment, I would make it up to her.

Letting myself out and off the hook by saying how I could tell she was busy, I made my way down the elevator to the first floor and asked the operator to hit the panic button to summons Carl Watson. When he arrived with his usual sour expression we went to the same private corner where we were the last time. We had a lot in common now.

"Your men are doing the job, Brandon," he said. "My man is young and inexperienced, has no idea that anyone is watching him."

"I'm here because of your man. I need to at least know what he looks like, since I'm all over this town trailing other people. If one of them leads me somewhere and he's there too I'll recognize him."

"Sure, he's working his shift now," Carl said. "You just want to look at him or do you want to go over there and meet him?"

"His name and a picture will do, but I need to visit the lab sometime when he isn't there. I've been to the lab at U T where Bob worked under the University Grant."

In a few minutes I related everything I had been doing to him and he also filled me in on a few things. We both needed all the help we could get and we had to trust each other. When I finished he had a comment.

"You would have had to sign your life away if you needed to get in that lab," he said. "The nanotechnology stuff is still a little if-y"

"Is the laptop at Gladstone like that?" I asked.

"No. We have had to take it somewhere a few times. It locks in a special box with a chain and handcuff attached. You wear the handcuff on the left wrist. The weight is not much and the chain is short enough to carry it swinging without it hitting the ground."

"What sort of places did it need to go to?"

"Special computer technology places to get a bug worked out of it, or somewhere Bob would take it to give a demonstration to other nanotechnology engineers."

It dawned on me that Bob might have cuffed it on to himself and just taken off with it. But Carl said he had to fill out a requisition and have it approved by the C.E.O. Harold Remkin. Carl believed it would have been impossible to get the laptop out of the lab without the cooperation of the guard who was on duty at the time. By the process of elimination you had to accept the fact that if Bob had walked off with the laptop it would have been last Friday at the time he was logged out by the guard. He had never been logged in on Monday morning so it had to have been Friday when he was logged off at 4:47 PM and the guard on duty was a young man by the name of George Campos. I wanted to look at the guard's station.

Carl took me for a short jaunt inside. It was on the same ground floor area we were already on. We took the corridor which extended between the switchboard console and the wall where we had been seated for our private meetings. About half way down it we came to a brick archway with an opening and stepped inside Carl's office. To the left there was a mahogany desk and chair and a few file cabinets. He went over to a file cabinet by his desk, opened it and removed one of the files. Then he unclipped a color photo that had been attached to the file, made a copy of it on his computer scanner and handed it to me.

"That's George Campos," he said "The best of my three men, but it had to be him."

"Only if your theory is right," I said.

"It sounds tough having to put in a requisition to Remkin, but sometimes he just called Harold on the cell phone, told him there was a bug and Harold said go ahead and do the requisition later. The guard at the station goes along if the laptop leaves, so you see George should have gone with Bob, but he didn't."

"What did George say about it?"

"He claimed Bob just checked out at 4:47 without the laptop and so he remained there till the end of his shift the way he was supposed to do. The entire week end one of the guards was at that station in front of the locked door to the lab. They don't have a key to it. Bob always locked up when he left and returned the key to Remkin's office."

Now that I had the whole picture, I thought maybe Carl was right. George could have been in on the conspiracy, when he signed Bob out Friday afternoon. For all I knew Bob could have been wearing the cuff attached to the laptop. There were too many possible scenarios. I could see it was driving Carl nuts.

CHAPTER 5

It was about noon and I thought I'd try to smooth over my fouled up relations with one of the star players on the Gladstone administration team. I took the elevator up to three and caught Marsha in the outer office away from her computer. I grabbed her hand the way she had done me on our first meeting, and I pulled her out the door to the elevator. She was laughing, on the verge of hysteria and loving every moment of it.

"Where are you taking me?" She said. "I think I have a previous engagement."

"Only thinking doesn't count." I flashed an old deputy badge I had in my wallet and said I was an officer of the law so she'd have to come along. She kept on laughing and half trying to pull away all the way down the elevator and out to the car.

There was a barbeque, cafeteria style restaurant I had noticed only a few miles down LH-1. We were both chuckling, while we put food from the steam table onto our trays.

"I didn't know you were such a buffoon she chortled.

"So now you know."

"Good, you're entitled to mix a little play with work."

"Can you mix it with your work too?"

"Yes, especially with my work," I answered. "Sometimes I think I'm the center of the universe and I need to lighten up."

Marsha told me I couldn't be the center of the universe because she already had the job. We might be able to get together now and then, while we allowed the universe to run itself. We could take a short vacation. The one we were on now was filled with beef and cheese enchiladas, mashed potatoes and corn on the cob. They weren't all things that went together, but we enjoyed it just the same. When we finished I put her back on the elevator at Gladstone and I put myself on the freeway headed north.

In many abduction cases the victim was held at a place very close to the spot where he had last been seen. It was safer for criminals to only move the victim a short distance. The perpetrators must have planned on where they would abduct Bob in relation to the holding place. I thought they had taken Bob from the parking lot at El Dorado where the contrived altercation had taken place. But that theory didn't work any better than Carl's. I had an explanation for how Bob was taken on Saturday night, but it didn't go along with the idea of the laptop theft. Why would Bob walk out from Gladstone Friday afternoon with the laptop cuffed to his wrist? He would not have been likely to drop off the laptop somewhere and then go on a dinner date. When the dinner date didn't go well he had dropped Marsha at her place, and then returned to the night spot for a beer drinking spree. That was not the act of a man who had just performed a robbery at his own work place the day before.

I turned off the freeway at Clover Lane and parked in front of the brownstone condo at 1971. I did the walk around thing to no avail. Only the birds and squirrels seemed to know I was around. I climbed back in the car and drove to the other end of the winding Clover Lane.

On the way back to the LH-1 freeway I drove the entire distance of each of the cross-streets, which had been given the names of trees. While on Maple Street I had to slow down and almost stop for some construction work being done. It was a blighted area with several condominiums seeming to be empty. I remembered it because on the other streets all the condos were filled. I supposed it was some problem with utilities on Maple that caused the vacancies.

The last tree street I investigated was Pine. It was the first one you came to from Clover Lane and the last one I checked before starting back to the freeway via Clover Lane. I had just pulled onto the access road when it hit me. I made a u-turn back onto Clover Lane and went to where I had seen an office sign in one of the windows. I parked in back of a red pick up and went in.

After going around between the saleswoman and several different maintenance people I finally got the manager who had all the keys. I convinced him that it was important for me to view the vacant condos on Maple Street. He saw it as a way to avoid a confrontation with the uniformed police, when I insisted that would be the only other option. He smiled and said for me to follow his pick up and he would open them up.

"I need a little more than that," I said. "This is a possible kidnapping case that could easily erupt into a Federal Investigation."

"What sort of information do you need," he said

"If there is anything suspicious going on in any of these condos tell me, now."

"No, these are high class, expensive condos, but vacant ones get broken into."

"What about the ones vacant on Maple Street?"

"The only thing was a report from the residents at the far end. They claimed there were noises coming from the vacant condo next to them, a thrashing around sound."

"That's what I'm investigating," I said. He motioned for me to follow his truck.

I followed the red pick up back to Maple Street and we went inside to look around. I pulled out drawers and looked in all the cabinets in the kitchen and it was just as clean as the rest of the house. There were marks where one of the windows had been jimmied and it was left unlocked. The manager was surprised to see it, when I pointed it out.

"How long has this been vacant?" I asked.

"They just finished moving out last Saturday afternoon."

The timing would have been right if Junior Gilbert and his friends knew it was empty, Saturday afternoon and had used it as a temporary holding place for Bob. It wasn't much of a stretch to think that, because things never go the way you think they will. The perpetrators may have planned to wait till Monday to do just as Carl thought they had done on the previous Friday afternoon. Carl's theory was wrong. They already had the laptop or were assured of being able to get it without Bob's help, so the plan was changed. Bob was taken to the condo on Maple Street until the more permanent place became available. Where it was and why they had to wait a day or two, I didn't have a clue.

It was the dangerous dusk time when I parked at Gladstone and went up to the third floor. I felt as if the wolf had followed me into the elevator. I looked behind me as I got off. This time it was Harold Remkin I was there to see. He took me farther down that back aisle than I had ever been and sat me down in his spacious office on the other side of his huge mahogany desk.

"Did you have something to report, Mr. Brandon?"

"No, but now, I'm informed well enough to ask better questions."

"Like what?"

"When Bob would finish his work for the day and return the key to you what happened to it?"

"I put it in the vault with petty cash."

"Where is the vault?

"It's easier to show you," he said with a courteous smile. He pushed the swivel chair away from the desk, got to his feet and led me to a small cupboard door. It was unlocked but when he opened it I could see the front of the vault with a combination lock. He said he was the only one who knew the combination and that it was not written down anywhere.

"What happened when Bob checked out of the lab and you weren't here?"

"That has never happened because he would always give me a call first."

"Are you telling me the key to the lab door he turned in every day was the only key?"

"No, there's always a spare key kept in there also."

Harold was turning red faced, not liking to be the recipient of all my careful scrutiny. When I asked him to open it so I could be sure there were two keys in it at the present time, I thought he might strike me. He just swallowed and worked the combination. He pulled the door open and we were both surprised to be looking at two slots with only one key.

"Don't blow a gasket over losing a key," I said.

"I would swear they were both there when I returned the key Friday afternoon."

He pulled out a few bills and some change. The inside of the vault was empty except for the single key. We bent over and examined the floor thoroughly in case one of the keys had fallen and bounced. I reassured him that it only proved he was human. He was still in charge here, although he was a bit tarnished. As I let myself out to the elevator he was talking to Marsha Brooks on the cell phone, asking her if she had any idea of what had happened to the other door key. I thought possibly he hadn't told me the truth. It wouldn't be the first time a CEO had shielded others on his team to take full responsibility. I didn't believe he was the only one to know the combination, but I let it go.

The same old ghostly panic followed me all the way down the elevator. When I got off, the cell phone went off in my pocket. I juggled it while going through the outside door. I answered it and listened to a male voice asking me if I was Dr Bob Clawson.

"You're on the wrong side of me already," I said. "It hacks me off every time I think of some pompous ass who's a PhD."

"Yeah, I'm sorry about that," the voice said. "This is Junior Gilbert. You told my brother you could help me get a job."

"When I say I'll do something I deliver."

"Yeah, I'm at the Hungry Jim Grocery at Smith Road and Ives Street."

"Know where it's at," I said.

"That Humanities class was a real pain. Sorry I got pissed off that time and it'll be good to see you again, Bob."

I said I felt the same way about that danged required Humanities class and hung up. As soon as I consulted my Map Book I did know where it was and I hit the road. Students who fill up their schedule with math and science classes tend to go to sleep in most of the school's required classes. I was glad I remembered that much about college. If this guy was OK I wanted to verify it and cross him off my list. I thought by the sound of things that it would be exactly what I would be doing. The grocery store was located on Ives Street not far from the drag. It all seemed to go together. Nothing strange to find the store closed and most of the outside lights burned out.

When I parked and got outside of the car at the Hungry Jim Grocery, the darkness was on me like thick fog. Junior hadn't said the store was closed, but then I didn't think to ask either.

The red-neck with the flat nose was sitting in the driver's seat of his brother's truck, and appeared to be alone as he should have been, waiting for Bob to come along and recognize him. I was not Bob and when I approached him he failed to pay much attention to me until I came right up to the open window. He put his long neck down in one of the beverage holders and opened the door all in one motion. Two men suddenly flanked me on either side while Junior worked on me from the front. It was a surprise trap that never should have worked, but it did.

I think one of them used a sap on me and I could feel the immediate loss of strength. I was forced to the floor in the back of the four door pick-up. The two in the back seat with me were using their heavy boots to great advantage on me, as Junior took over the wheel and drove somewhere. I was never able to get up from the floor to see where we were going. When we arrived at the place, I was carried inside while unconscious and didn't see anything. However, I had recognized George Campos as one of the men.

The rest of the night and most of the next day I was in and out of consciousness, only hearing bits and pieces of conversations, but it all seemed far away and meaningless. The room was kept dark and someone had tied my hands together behind me and also tied my feet together, bent my legs up in back of me to meet the hands and tied me all together like a roped calf. I was on my side on the floor of a medium sized room. There was no furniture so I could flounder around all I wanted. The trouble was that I was too weak to get the ropes loose enough to do any good. I'd try it a few times and get so tired I'd just go back to sleep until I got rested enough to try it again. Junior and the man I could not identify came in the room one time to check on me. They didn't talk at that time, but when they went out they left the door ajar and I was able to make out some of their conversation. Junior said, "That grant should have been mine, Tom." The other man agreed quite earnestly, "Yes, it certainly should have been."

As the sunlight started drifting through the drawn blind of the only window, I started working harder on the ropes. Rolling from my side to a wrestler's bridge on my stomach, I even tightened my abdomen and rocked back and forth a few times. It didn't do much good and I flipped back on my side again. I seemed to do better just pushing my hands and arms as hard as I could in a motion away from my back. Doing the same with my feet and legs started giving me some wiggle room. I remembered something I had read about how the greatest escape artist of all time had wiggled his hands and wrists to free them. I concentrated on my hands for a while and thought the rope was getting a little loose, and that gave me the incentive to force my hands and wrists even harder. If I could only get those hands free I could more easily untie the rest of that mess of rope. It didn't matter that they were getting all bloody. The blood would act as a lubricant and maybe even speed up the process that would lead to my freedom.

I thought I had heard the sound of footsteps and it put me into a terrible frenzy of activity that resulted in my hands coming free. The rest of the rope came untied quickly as I had imagined it would, and I stretched out that fearful surge of energy through the open door to a kitchen area. There were two small boys tossing a ball I could see through the kitchen window, probably the sound I had mistaken for approaching footsteps. The dead bolt was locked on the exit door next to the window. Unlocking the dead bolt with one bloody hand and the door handle with the other I wasted no time getting out of there. I was still too weak for another fight. Even one of them would have been enough to have finished me off.

The two boys tossing the ball froze in mid-toss when they saw me coming through the door. I could see my bloody and bruised body mirrored in their wide, horror stricken eyes. When I reached in my pocket to find the cell phone, the boys must have thought I had just escaped from the law and was going for a gun. They were gone so fast I didn't have time to talk to them about my predicament.

I drug myself, more than walked, to the shade of a large tree and leaned against the trunk. Then it came to me that I was on Maple Street. Junior and his friends had not been so innovative. They had taken me to the other vacant condo at Four Leaf Condominiums. One of them had driven Bob's car to his condo and they knew about the vacant ones on Maple Street. Their kind could always jimmy a window and get in.

A loose water hose was attached to a wall hydrant. I turned it on and sprinkled myself enough to clean off the dried blood. Soaking my bruises in the cold water felt good and I was invigorated enough to hide in the wooded area and use the cell. I called the switchboard operator at Gladstone and asked for Carl Watson. He wasn't available and the operator put me through to Harold Remkin. I gave him a full report of my miss-adventure and said they should hold George Campos if he was foolish enough to show up there. He said the company would pay for me to stay at the Crystal Palace Hotel while I recuperated. I told him the police should be sent over to the Condo where I was held to pick up those guys if they returned, or at least have Carl go over with one of his friends from the CPD. I gave him the address of the Condo and he mumbled something. I told him I'd call a yellow cab to take me over to the Hungry Jim where my car was parked, and then I could drive it over to my luxury hotel at the lake.

When I got there I was met by the grounds keeper who showed me to my room and gave me the key.

CHAPTER 6

After finishing breakfast I stepped outside and was impressed by my surroundings. The hotel was located on the edge of the lake with several storage buildings and a mock-up lighthouse on an acre of land. I got it from the grounds man that the almost full-sized lighthouse had been placed there by a local movie company. It had been used in the making of a documentary film. The owner, Albert Hoffman, had purchased it from the film company to remain on the property for some commercial notoriety, and it had become a well known landmark. It was sunny and about seventy degrees with a nice breeze, when I started down the gravel path. It led down to about twenty five yards to the right of the lighthouse. I went that far and watched a young couple with two children walking at the end of the path on a small boat dock on Lake Crystal. The father went off the dock into the grass and down by the water. He started skipping rocks out into the water. The two boys joined him and started to mimic his behavior with their own flat rocks.

My legs were still a little shaky and besides, I didn't want to invade any of those precious family moments. I turned around and went back to my room inside the hotel. I poured the remainder of the coffee from the urn and savored the special blend. I'm sure I had slept at least ten hours. The rest had done my body some good and I was starting to mend. I leaned back in the reclining chair, picked up the remote and turned on the news with the volume down low.

The cell phone went off and it was Marsha, wanting to know if she could do anything for me. Harold had filled her in on my entire sordid story. It sounded like he had assigned her to the job of keeping up with me. Since I had stirred up things and caused all the commotion, they wanted more detailed information regarding my plans for the future on this case. I gave her directions to my backroom office at Southland Realty. They keep my extra key there for emergencies like this one, so I called and said Marsha would be coming with a list.

I went out to the car and got my light-weight portable typewriter and some paper, so I could type my report to Harold, about what I had done and was planning to do to solve this case. Carrying my tools and materials over to sit on the end table next to the recliner, I put the typewriter on my lap, leaned back in the chair and started typing the report. It was what I should be doing, a sitting job that would get me off the streets for a while. There was an excellent view of the Lake from the window and as I enjoyed it Albert Hoffman walked into view. He caught my eye and waved. In a few moments he rapped on the door, and pushed it open.

"Stay where you are," he said. "Got everything you need here?"

"Yes fine, or will have soon. Marsha is bringing me some things."

"I've known her since she was a tadpole." He smiled reminiscently, an elder that should be retired, but would more likely die on the job. Not that he needed to work. He had deep roots in the community, and his strong masculine features were precursors of his vitality. He had piercing blue-green eyes a pug nose and natural reddish-brown hair. I wondered if he owned many shares of Gladstone stock. A lot of the locals did, since they had learned about the super-laptop.

"Do you know a lot of them over at Gladstone?" I asked.

"I see Harold and Von every week at the Rotarian luncheon," he said "We own a boat together down on the coast and go deer hunting together." That told me that if he had some stock he would hold on to it, but probably not be buying any more for a while.

"Well, I have to make my rounds now. Give my regards to Marsha." He went out and the door slammed. I went back to typing my report. Albert had failed to make a comment about the apparent beating I had taken. He must have known about the business I was in from Harold and decided to keep his mouth shut. I appreciated it and returned the favor by not quizzing him any further.

Reluctantly, I was back to the report and not doing too well at it. I was still working on the first part, which was supposed to be about what I had accomplished and my observations regarding the people I had interviewed. It seemed as if all I had done was get beat up, and the only one I had given a complete interview was a dippy piano player. Thinking about my future plans was also a bust, so I just typed some general stuff about how I intended to seek more information about Bob. I also mentioned how I would have Carl Watson bring to bear all the forces at his disposal to find and arrest George Campos. It was not a very long report but the best I could do, and Marsha would have it ready to take back to Gladstone.

As I took the paper out of the typewriter and placed in on the table I was rewarded by a slight sense of euphoria and the comfort I enjoyed leaning back in the recliner. It was about as good as being in bed. I realized that it was getting hard to keep my eyes open.

The door must have been unlocked and all of a sudden I was on my feet being led and half carried to the bed. "Come on now, just a little more," Marsha was saying. I dropped my upper body on the bed and she scooped up my feet to fit on also. I rolled in some to make room for her to sit on the side of the bed, which she did. She propped up the pillows for me, and we carefully embraced and found a delicate kiss on the mouth that turned ravenous. She pulled back, afraid she was hurting me.

"I need to let you alone now," she said. "Dan, you look bad and you need to rest."

"Yes, I'm going to. The report for Harold is there on the table."

She walked over and took a look at it. "This is fine, Dan."

"Albert said to send you his regards."

"He's my Dad's old hunting buddy, known him all my life."

"You must have found everything." My canvas bag was sitting by the door. She pushed it to the table and made room for it by taking off the typewriter and a few other things, which she sat on the floor.

"There's just one thing I need from the bag." I said.

"I bet I know what it is." She put the bag on the table, opened it up and took out the thirty-eight Smith and Wesson along with a box of shells and brought them to me. I opened the box and dumped a few shells on the bed. I broke the barrel of the revolver and filled the cylinder with shells. "Thank you for heeding my advice not to load it," I said.

"If I had shot myself you would have never forgiven yourself."

"No, I wouldn't have been able to, and thanks again." I shoved the revolver under the pillow for the time being and pretended to be ready to go back to sleep. If we had done what we almost did, when both of us were in bed, it would have been a problem for me. I was never good at juggling women around. Marsha didn't even know that Claire existed and under the circumstances I thought it better to keep it that way.

"It's good you have the gun to protect yourself, Dan, since those guys know you could identify them."

"I'm a light sleeper and that helps," I said. "Lock the door on your way out and be sure you call me first, when you come back. Don't just pop in the door."

"Oh, no I won't be popping in," she smiled. "Don't be leaving the door unlocked."

We were sounding like a married couple. It made me think I needed a break from this relationship. "Got the report?" I could see it perched on top of the typewriter.

She scooped it up and went through the door, locking it. I thought she might have needed a break from me too.

After I emptied the rest of the things from the bag and put them in drawers or on hangers, I took a shower and gave myself a half-shave with the electric razor. I didn't have any concussion and the hickeys would soon be gone. I put on a clean pair of Dockers and one of my favorite Safari shirts with a lot of pockets. The shirt was made to be worn outside and cover my holstered Smith and Wesson. I had on the garb and wiped off the white canvas walking shoes I had been wearing. I put them on and laced them snug but not too tight.

The sun had reached its zenith when I parked at the University hole in the wall place and went through the tunnel with Wallace Norton. I had informed him of my need to learn more about what went on in the lab and he said I would need to talk to Nelda Whittington. When Nelda came out of the lab to the vending machine area he left us and left the key to the tunnel door with her. He gave a short explanation about a graduate class he had to teach and how Nelda could let me out when I was ready to go.

"Wallace said you had some trouble and I see you've been in a fight," Nelda said.

"I'm fine, and I don't have time to go into that now." The cap I had been wearing had covered the spots on my head, but I had instinctively taken it off when I came inside. I put the cap back on again so she would forget about it.

"I understand you need to know more about our lab here." Nelda said.

"Yes, as I learned more about the research lab at Gladstone I was made more aware of the differences between the two research locations. This one seems to be more of an environmental risk if I understand it right."

"We don't feel that way about it, because we use all of the protective measures here. I can't talk about the details of things to anyone unless there is a need to know. Can you be more specific about what you need to know?"

"I need to just get a working knowledge of how the process worked between this lab and the one at Gladstone," I said. "The least technical the explanation, the easier it will be for me to understand."

"Well, I'll start with where we left when you were here before, Nelda said. Do you remember I said as we made improvements here and tested them we moved them over to Gladstone?"

"Yes, and I've been thinking this process can have some bearing on how the robbery took place."

"When the module here at our lab was improved and made safe I would tell Bob about it and he brought the laptop from Gladstone over here. We would move the new module from our laptop to the one from Gladstone. You see we had two laptops, one here and one at Gladstone."

"Do you still have one here?"

"Yes, but without Bob there isn't much going on with it." Nelda said.

"Does it have the yottabyte module in it?" I asked.

"Yes, but the highest level of the module was on the laptop at Gladstone."

What I wondered was if the people who had the laptop, containing the highest level of the module really knew what they had. Even if they made a techno-slave out of Bob and could count on him to work with them would that be enough? Did they need to continue with the same sort of arrangement with the University lab to insure continued success?

"Has anyone contacted you about the research here on your laptop?" I asked.

"No, you mean since the robbery of the one at Gladstone, I assume."

"You see what I'm driving at?"

"You're thinking that the perpetrators might want some sort of technical relationship with our lab to keep things going with some other computer company."

"That was exactly what I was thinking," I said. "Those people know the value of what they have, and they also know they can find another company CEO who will make a deal with them if they can present things as being foolproof and legal."

"They would have to depend on Bob to help them. He would never do that. They're just holding him as a prisoner."

"I've heard of mind changing drugs and big time trickery being used. The people running the show have to be well educated and intelligent. I think Junior Gilbert was a cat's paw. I know he was jealous of Bob because Bob received the grant from the University at Austin. Junior and his pals were just hired help."

"If the guys who beat you up were being used by a more sophisticated type of criminal, that would mean they may not even have the super-laptop. How will you find out where it is?"

"I find the cat's paw and follow him to the one who has it."

"That might work, Dan."

I thought Campos had to have been a conspirator in the laptop heist, but I couldn't be sure of it. No evidence, so it was all circumstantial. That lost key could have been found by anybody. Now that Campos was no longer around to be followed I decided to give Willie a call and have him help me to track down the others. I put him onto Junior, giving him the address where I had visited Junior's older brother, and then I got back to networking with Nelda.

"What about the nanotech design in the super-laptop module, Nelda?" I asked. "Does the module need any attention to preserve it, like being packed in dry ice?"

She said it shouldn't need any special treatment, with an injured look. I thought it might be a good time to leave, so I said it was good to see her again. She seemed eager to go back to her lab. That was when I noticed her hair. It had been short and straight before. Now it had some curl in it, more stylish and longer. When I came back the next time I'd have to compliment her about it. Something happened to her to cut through the depression and renew her spirit. I wondered if she had discovered the place where Bob was being held and been with him for a while.

CHAPTER 7

It was only a few blocks to the drag for me to get lunch and visit a place of special interest to me. I drove to the El Dorado and parked in a parallel space in front. It was the slow time, before Marvin came in to take charge of the piano bar. A short, stocky man with a long chin and a bad haircut was tending the bar along the rear wall. I sat down on one of the stools and ordered a cheeseburger with coffee. He brought it to me in a few minutes and we started talking. His name was Tim Larson. He had been the one who worked with the manager that night, and broke up the early morning fight in the parking lot. I hadn't wanted to evoke any sympathy, so I had left my cap on to cover my head wounds. Tim was observant enough to notice a slight mouse under my left eye, and asked if I had been in a fight. I told him how I had set up a trap for Junior and fallen victim to it myself. He didn't think that was funny.

"I can see how that could happen," He said. "Those three were together last Saturday when we had that trouble."

"Do you know if they have any records I could look up with the police?"

"That Thomas Oliver has been up for armed robbery. He was circling around trying to flank Bob, while he was getting into it with Junior. When I got it broken up Bob got into his Chevy and those three got into a pick up and threw some gravel going after him. I'm not surprised to hear they may have caught him."

Business started to pick up and Tim went to serve another customer. I called Willie Harland on my cell phone and asked him if he was getting along with Watson. He said he was getting along, but he didn't have to worry about tailing Campos any more. The young security guard had been found dead with one bullet to the head and one to the body.

"Where did they find him, Willie?" I asked.

"A trucker spotted the body in high grass at the side of the road on Highway 71, in Bastrop County."

"You stay with Carl, and have him use his contacts with the police to find a guy by the name of Thomas Oliver, gone down for armed robbery. When you find him call me right away, carry your piece and stay right on him till I get there. I hope he will lead me to someone else."

"He won't see me."

"Right, keep Dartmouth with you, set up a stake-out and I'll see you later."

Junior Gilbert would be long gone, but I thought I might find him again through his brother. I got a coffee for the road, went outside to my car and drove to the dirty-white frame house down the street. It looked empty, not even any front curtains. I drove a little past it, parked and got out. They must have flown out of there without notifying the landlord. The front door was standing open. I went inside through all the rooms looking carefully for any sign of where they might have gone, but there was none. They were smart enough to have cleaned things up, leaving nothing. Junior must have told his brother that he was in big trouble and needed to hide out for a while. I went back to the car and drove the short distance to Gladstone.

Watson took me back to his office and we sat down in his uncomfortable hard-back chairs. I wanted to know how he had been doing with his connection to local law enforcement, regarding the trace on Thomas Oliver. His friend, a detective with the Crystalville Police Department had contacted Oliver's parole officer and had an address in

Crystalville for him. Willie and Ray were given the info on him and were already setting up the new surveillance location. Carl had a few questions for me too.

"You still hearing bells in your head when you typed that report to Remkin?" He said. "He showed it to me, and it was garbage."

"Did you tell him that?"

"No, you kidding, I can't start up anything with him, but I think you could tell me a few more facts than that."

"Sure I will, just give me a little time. I wasn't thinking to well then, so I just typed up some general stuff for Remkin. What I did to stir things up was put pressure on Junior Gilbert. It precipitated the attack on me and put us to where we are now."

"Don't get me wrong, Brandon. I appreciate what you did and I think it's going to pay off."

I said that reminded me of my next plan and I filled him in about all the things I had learned from Nelda Whittington regarding the other super-laptop, which she had under her care at the University lab. I also told him how we had both agreed that there was a more sophisticated leadership behind the heist of the Gladstone super-laptop than the three stooges. One of them was dead and another one missing, so we couldn't afford to drop the ball with Oliver.

"I take it you want to join your men on the stake-out," Carl said.

"Yes, and have the CPD ready if we need to call for back up."

He assured me that I would have the backing of his entire team, which included the national police network. As far as we knew at that time, the FBI was not involved in the case. Most cops would be very happy to work through their entire career without ever having to work a case with the FBI. We were no different from the rest, but we were aware of the distinct possibility that the FEDS might come on our case at any time they decided it was to their own interest. They didn't have to give us a warning.

There was a late-afternoon wind twisting around outside. If it didn't turn into a full tornado, you always had a little breeze to help cool down the hot spells in July and August. It was my intention to be finished and out of Texas before that time, be playing around the trout streams in Colorado, without a care. I was still mulling that dream around as I drove onto LH-1, and into the inner core of Crystalville. I went to the low rent district, where the last hold out of the three criminals lived in contempt of the authorities.

My instructions pertaining to the location of the stake-out were to pull off the freeway at the next exit past a stretch of flat land with railroad tracks running parallel to the highway. I saw it and turned off. A gravel road ran along between the railroad tracks and the highway. I had been on it only a few moments before I came to a wooded area and parked at the edge of an open field. There was a ranch-house with several vehicles parked in front of it. The rancher had given the police permission to set up the stake-out on his property. I didn't see anyone around, but I could see the rented, blue van parked parallel to the tracks on the far side of the field. I walked over to it and the side door popped open. Willie motioned for me to come inside the van and I did. He had placed a bench seat in front of the side windows facing the tracks, and was sitting there using binoculars to view a vast space beyond the nearby tracks. He handed them to me and I sat down by the window next to him and adjusted them for my own use.

"You see the next road over there with the shack set back a ways?"

"I can see the road," I said. "But I don't see any shack."

"You have to train them in better."

I thought I understood why Willie could always observe, without being observed himself. I adjusted the wheels some more and still didn't see it. He took the field glasses away from me and made a quick adjustment, holding it for me while I looked. This time I did see it. It looked more like a small storage shed and I couldn't think it was large enough for anybody to be living in. He helped me again and I saw the vehicle parked to the left of it.

"It's too far, Willie, for me to make it over to that road before he's gone."

"I've done it," he said. "The guy drove off to get something at a shopping center and I had to follow him. Ray had to do it on his shift too." I couldn't argue with success, especially the two against one kind. I had worked for Willy to get started in this business. He was my mentor and had married my partner, Helen Wilson, from my Austin Police Detective days. She had quit the APD to join his agency in San Antonio. The next thing I heard about them they were married. He was at least ten years older but they got along well. It seems that women often go for the Willie type. He could always make you laugh, and he cared about people. In the looks department he was about average, going bald on top, with a well-proportioned thick nose that had probably never been broken. He wore his dark-brown hair long enough to about cover the bald spot, and was about six feet and two hundred pounds, with a dark-skinned weathered look.

"Twelve hour shifts are too long," I said. "I meant to get with you sooner."

"Looks like you had good reason," he smiled.

Instead of saying I had been playing some sand-lot football, I filled him in on all of the foul treatment I had received from the three thugs. He already knew most of it anyway, but he listened and was sympathetic.

"Sorry, I won't be able to get any pay back from Campos now," I said. "Might, just take his share out on Oliver since he was the one to use a sap on me."

"You mean if the same one that iced Campos doesn't do the same to him before we get around to him."

"That's a distinct possibility," I said. "And for all we know Junior Gilbert may also be laid out in tall grass somewhere with two slugs in him."

Willie nodded in agreement and reached for the binoculars. I held onto them and suggested he at least take a break if he wasn't ready to start his first eight hour shift. He claimed he was fine, but did take a break long enough to return to the local motel room which was only a few miles from the surveillance location. With the idea of talking things over with Ray Dartmouth, who was our youngest member, Willie jaunted over to his own vehicle and drove away. He had parked by the other vehicles by the ranch house to draw less attention to himself. I should have pulled into his space as soon as he pulled out, but instead I glanced through the long range glasses and thought I saw something move by the shed. I fiddled with the adjustment wheel and could make out the fuzzy outline of a car moving slowly away from the unimproved road toward the highway. The unexpected was always the most likely thing to happen.

Taking the binoculars with me, I ran to my car and raced onto the road throwing pea gravel all the way back to the ranch house. When I reached the highway I floored the accelerator and hoped I was on it sooner than Oliver, so as to get a little start on him. I thought that was how Willie and Ray must have done it. I could see the blurred object in motion was getting on the highway then, and I continued at top speed. The fuzzy outline cleared up and I drove with my left hand while I adjusted the binoculars with fingers and thumb of the right hand. The vehicle was an RV, maybe a Ford Explorer. In a few moments I was able to manipulate the glasses well enough to read the license tag, and I let up on the accelerator. I was still gaining on him, but we were in a little traffic now and I had to be sure not to loose him. I pulled up to a distance of about four cars between us. I tailed him out of Crystalville going south east. He got on Highway 620 and continued south into heavy traffic and I didn't want to risk loosing him so I came up to three car lengths for a while, but when he got on the Bastrop Highway I fell back to four cars and didn't think he was aware of the tail at any time.

I settled into the pocket of traffic on State Highway 71, thinking we might be going to Bastrop, but he pulled off at the Austin International Airport and I followed him into the large parking garage where he parked and got out. I drove past his parking spot on a little higher ground so I could keep him in my vision and see where he went into the large building. There was no hurry and I couldn't get close enough for him to recognize me. I thought he must have been looking for someone coming in on the daily commuter flight from Dallas. It was that time of day, so I just waited for them to return to Oliver's car. A tall, slender man, dressed in a black sport coat, light colored sport shirt and khaki pants came out of the terminal building and went to a short waiting area for a yellow cab. Oliver came out about that time and headed straight for his car. He was alone and I wondered about it for a few moments till I figured it out. Oliver jumped in his car at about the same time the man got in the yellow cab. He followed the cab and I followed him, taking it easy again. I had a good idea about where they were going.

Following the follower, I ended up at the convention center in the downtown area of Austin on Caesar Chavez Street, or First Street. There was a special computer world-wide-high-tech convention going on there all this week. It would be a good place for someone to peddle a super-laptop computer with a special module in it that would make the old computer technology obsolete, and the new nanotech computer in vogue. It was survival of the fittest in the largest sense of the word.

The nanotechnology had already broken ground with huge advancements promised in the field of Medical Science. All the leaders in computer technology from all over the world would be here. If Bob Clauson was half as good as I had heard, as far as being obsessed with the compulsion of building the fantastic small laptop, then certainly it would follow that the new explosion of information provided by a nanometer of memory space could bring about mammoth societal changes.

CHAPTER 8

Trying to learn all I could about the nanotechnology aspect of this convention, I read some literature about it that I found on a table in front of the entrance to the large auditorium. There was a lot going on in the way of presentations and exhibits for the next few days. I found a remote spot outside and made a call on my cell to Willie at his motel room. We had a short laugh about how I went to relieve him and then left abruptly taking along his binoculars. I spoke rapidly, as someone could come along at any time and hear me. Willie and Ray would be joining me soon. I said for them to meet me at the general assembly in the auditorium. I would be waiting for them in the entrance hallway, just outside the auditorium. I terminated the call and returned to the auditorium.

I had followed Oliver to the auditorium, and inside, where he was still seated in the rear area a few rows behind the man he had followed from the airport. Neither man had changed seats, when I returned to wait for my reinforcements. It really didn't matter if Oliver saw me or not, now. Willie and Ray had both seen his police mug shot, and Willie would be watching him. Ray would follow the other man to see where he stayed and anything else he could find out about him. I returned to the entrance corridor.

They arrived much sooner than I had expected.

"I wasn't going to say anything about speeding," I said. "But if they catch you around here, the case could be over before you get out of jail." They liked to make examples of people here. A few months ago the newspaper carried a story about a uniformed officer who tried to discipline a woman about wearing her seat belt, and ended up putting her in jail.

"I better get in there, so I'll be ready to stick with Oliver if he moves too far," Willie said as he opened the door. I could hear the speaker saying something about faster computers that would hold a lot more information. The door closed and shut off the sound from inside. I had dissuaded Ray from cracking wise and he was quite attentive.

"The way to play it, Ray, is to stay away from Willie so it looks like you're two stray guys looking for a seat, or some old friends in the audience. Whoever Oliver follows will be your man. He'll be a middle-aged, tall, blond, wearing a black sports jacket and kaki pants. I think he had on a white sport shirt with blue stripes, under the jacket. At some time Oliver will approach him, and if you can, place yourself so you will hear their conversation."

When I asked for questions, he didn't have any and decided he was ready to go into the auditorium. In the few seconds before the heavy-door closed again, I heard a different voice coming from the speaker's podium but couldn't make out the words. At another conference I had attended, the speaker cut off the lights and turned on a movie projector to do his talking for him. In no time at all a loud snore wiped out the sound track and the projector was turned off. My short presentation to Ray had been no more involved than what you would expect to say to a new member of your team. He was younger and a rooky detective, but you still had to be careful not to over-saturate someone. On the other hand if you didn't give them all the facts they could fail, and that would leave you feeling responsible for the failure.

My role would be to rove around the place without tipping Oliver to the fact that I was with Willie or Ray. I didn't care if he saw me and in a way I did care. I would prefer he did not see me at all, because I didn't want to curb his operation. I did not want him to run because he thought I was going to arrest him. I still wanted him to take me to his leader. I made sure my cell phone was on in case my men needed to get in touch with me and I needed to be patient. You should never rush something like this.

While I was toying with the cell, it rang and I answered it. Claire was only a few blocks away at her apartment. She had put in a full day, but still had some work to prepare for court tomorrow. When I told her I was nearby at the Conference on Global Technology, she said we could meet me at the Liberty Café a coffee shop that adjoined the Center. We had both had two meals and were not hungry, just wanted to talk and have coffee. She arrived with the lovely Irish smile she wore most of the time. Her mother was Irish and her father a Frenchman. It seemed like a good combination to me. Claire was very reasonable but was also a fun lover and could be quite sexy at times.

"Oh, I see you got the vegan chocolate-chip cookies I like," she said. "And the fresh pot of coffee's here." She poured us both a cup and took a cookie from the cut glass plate that held more of them than we were likely to eat. I took one too and fixed my coffee. She was wearing a colorful peasant skirt with an orange blouse. She wore very little make-up and never really needed any. Her hair was reddish-brown, shoulder length with a light curl. She had an ample forehead and an elegant nose. Whether she was walking, standing or sitting it was like there was a book balanced on her head and yet she always seemed relaxed.

"Been playing that sand lot football again?" She asked.

"Yes, I always get to be quarterback, but I take too long passing and I get clobbered." She never made too much of things and like now, I always appreciated it.

"So, how's the Global Technology Conference going?"

"They talk a lot about nanotech terms like nanometers but not nanobytes.

"They have an inside joke that they can't have one, so a new unit is used."

"Well, I guess by the time I finish this case Gladstone will be able to pay me." I laughed without knowing what I was laughing at.

"That's right" Claire smiled. "I wouldn't worry about it. They're sure to pay you, no matter what."

Claire finished a chocolate chip cookie and polished off the cup of java. I poured us both another cup and tried one of the cookies myself. The endorphins started to kick in from the chocolate and the coffee. I was well on my way to forgetting about the memory units of modern computers. It was better to know there were tricky things about technology, though.

"So, Willie is here working on your case, now."

"It's a little slow in San Antonio, but that's how it goes in our business," I said. "Sometimes when I needed a case I went to work for him."

"You know Dan, it seems like your case is so big that the President might send the National Guard on it."

"I'm starting to feel that way myself." How every time someone looked at me a little strange, I thought it was the FEDS. I felt like maybe the President did know about this tricky technology at Gladstone. There were huge political implications involved in this tricky technology.

"I was just being facetious, Dan, but this Global Technology meeting is really an event. There are a lot of turbans around here, people with power and money, who are looking for more of the same. I hope you're being very careful."

Claire was quite an asset to me, like a wife, and I had offered to marry her. She said that we were not ready for marriage and was probably right. I had left it at that. It would be better for her not to get any more involved with me than she already was. I thought I might have told her too much about this case. I had been selfish and wanted her opinion. Without really thinking about it, I used her. She had said that sometimes she used me too, and that was what a relationship was about.

"I always appreciate your input, Claire, but I don't want you to worry about me."

"I should be careful where I put my input," she smiled.

We had a laugh, and I poured more coffee. We both had another cookie and changed the subject to something more suitable, like how bout them Long Horns. We both agreed that they were doing very well as they usually did.

It was good for me that we were the type to get high on chocolate chip cookies and coffee. The caffeine hyped up thought and conversation. If we had to have some joints or heavy liquor every time we met I couldn't keep up with what I do. She would have trouble with her legal career too. We loved what we did for a living. We did have a little wine at the end of the day.

"You have your key, don't you?"

"Yes I do."

"I'm going back now. I have some more preparation to do for my case. Say I do that, you check things around here till you get to a good quitting place, and then come back to the house."

When she was gone I walked across the street, where I had seen Oliver's car parked in the Hotel parking lot. As all criminals he was looking for one large score, so he could retire to the islands. I checked at the desk and found he was checked in to room 403. Thinking I'd be ready to investigate some rooms, I talked to hotel security and set things up. Then I crossed over Caesar Chavez Street to my car, and drove to the anomalous downtown apartment in about five minutes.

Claire had just finished her preparation. We drank a glass of wine and went to bed. It was a lot easier to get up in the morning than it had been for several days, since my mishap. After I had the breakfast tacos and home fries that Claire fixed, I was up to nearly a hundred per cent. She gave me a coffee for the road, in a plastic cup, and I drove back to the Convention Center and parked in a new space at the coffee shop. I left the half-full plastic cup in the car and went inside. The first thing I did was to buy one cup of coffee and place it at a vacant table by the back wall. I sat down at that table in one of the matching dark-oak chairs and called Willie's cell phone number. He answered on the third ring and sounded like he had just been wakened, but didn't want to admit it.

"I said I'd call if I needed back up, Dan," he said. "We don't know much yet, except the name of the man Oliver was following."

"That's good, Willy," I said.

"Hold on a minute . . . here it is last name first. It's Denison and first name is Calvin. He goes by Cal. Ray overheard a little of their conversation, when Oliver finally made contact at the big-time-hotel-bar across the street. They're both staying there."

"Did you have time to get it all from Ray, yet?'

"I got it all over the cell phone about the time the party ended at the bar. We have our own rooms, and only talk on our cells. Oliver is acting as a go between. He didn't give Denison a name, just a place where he could look at the super-laptop. Old Cal is supposed to meet Oliver at his car here in the hotel parking lot at eight-thirty, that's less than an hour from now. Oliver said he had his instructions as to the route he was to drive. They're supposed to be intercepted somewhere and Cal gets to see the product. Ray is going to check their rooms while their gone. We have permission from the Hotel, and I'm ready to go out the door, and get set up to follow them. What else can we do?"

"Nothing, you've handled it perfectly. I'm not going out on the road, and take a chance of being seen by whoever meets them. That's who I want to sneak up on. I'm going to help Ray search the rooms. What room is Denison in?"

"He's in 504."

"Tell Ray to call my number, let it ring twice and hang up to let me know they're gone. If anyone can pull off a tail job like that, it's you, Willie."

"OK. I wish us all luck."

"Yeah, luck more than anything," I said.

Ray was waiting at 504 with the Hotel Security man, who unlocked the room and stationed himself outside while we went inside to start our search. I thought Dartmouth looked even younger than he was with his eyes wide and his round face tense. It was better to be that way than be self assured with the phony over confidence that can be seen in a lot of young people. For someone no more than twenty five, he was doing a great job. I suggested he go through the clothes on hangers and in the chest of drawers while I went through the luggage.

I found a much better than average assortment of credit cards. All of them were gold or platinum. It was a mistake to leave them in the room, but people have a lot of different ideas about the correct way to do such things. He probably paid cash and had the cards only for an emergency. If he lost his wallet and the cash, he would still have the cards in the room. Those who travel a lot are advised to carry enough cash to get food and other necessities in case of a power outage caused by some terrorist act. Cal was prepared. I give him that.

My next find was a brochure about a Dallas computer manufacturing company by the name of Comp-ease Computers. I also found some more literature from the same company with a short list of management personnel. There was a Calvin Denison listed as Head of Production and R&D. It sounded like he had a good job with a reliable company. I had heard of Comp-ease before and thought it was a reputable company. But I wondered how Tom Oliver knew enough about him to be at the Austin Bergstrom Airport waiting for him to arrive here. I had to assume that Oliver had at least seen a picture of him or been given a very good description of him. Oliver had refused to be cut out of the action, and we had to continue to give him enough rope to hang himself, and to hang the pseudo-intellectuals who I believed were really behind this criminal venture.

"I found some notes and literature about the conference and that's it," Ray said.

I had only a few moments to glance at what he found, and my cell went off. We got outside, so the hotel man could lock up. It was Willie calling from a spot near Luling. He needed back up.

He told us how to get to the place where he was holed up. It was in South Central Texas, about forty miles south taking Highway 183 South. I thought we might get there in time to make an arrest, so I let Ray kick the accelerator pedal. If we had a siren I'd have put it on. We were in Luling in not much more than a half hour.

It was ironic that we were investigating a high tech crime in this traditional oil town. Some of the people here had made a fortune pumping oil for many years, but by now most of them had diverted that money into clean industries. And even these were under specific regulations. The electronics industry recycled materials from old out-dated equipment and had to pay heavy fines when they were caught dumping in land fills. Our national government had spent the last two decades working to improve these regulations. It had been a must do situation.

CHAPTER 9

Ray had his car parked in the hotel parking lot, as it should have been. Since mine was still parked at the coffee shop, we took his car. He drove fast and I made no complaints about the speed. It was still pretty early in the morning for all this intrigue, but you have to go with the flow. The traffic was starting to build up and the flow slowed down. We were on state highway 183 approaching Luling from the north, when we came through a patch of oil country and saw only one of the wells pumping in celebration. There was still the need for lubrication without combustion of gasoline engines, but they liked to remember the old days and how they got rich to begin with.

There were bright-colored silhouette figures attached to the horizontal members of each crankshaft on the wells. One of the jig saw cut outs was a large bosomed woman eating a big crescent shaped slice of melon. It oscillated vertically with the rhythm of the crankshaft. Another well had a cowboy riding a bucking bronco, and still another had a giant butterfly with bright blue and yellow wings.

The butterfly was the land mark Willie had given us. We turned on the next road to the left and Ray hit the brake, when we saw Willie's car parked off to the right side of the road. He stood in front of it and motioned for us to park. Ray pulled off, and we got out. I didn't like the stench of oil, but the production had been greatly decreased. A lot of the oil people had got the word by now and were investing heavily in nanotechnology.

"I'm close enough already," Willie said. "Denison walked in back of that house with the Mexican tile roof. He came back a few minutes ago and got back in the car with Oliver."

"Willie, above all else," I said. "You go when Oliver goes and stay with him. We'll check the house and whatever's behind it."

The dark explorer type vehicle backed out of the driveway, turned around and came toward us. Willie opened his trunk and walked over to the tire farthest from the road in the rear. He kneeled down like he was working on the tire. I ducked down so Oliver wouldn't see me, as Ray pulled back on the road and drove toward the house. I looked in the rear view mirror and saw Willie go after the RV keeping a respectable distance behind it.

Ray tromped the accelerator and we went almost airborne to leap onto the driveway of the gray-brick house with the Mexican tile roof. We jumped out and immediately noticed that the house was empty. We ran through the yard to the rear with guns drawn. There was a big drop off down to an alley. A blue BMW drove out of sight at the other end of the alley. There was no way for Ray to drive from the yard onto the unimproved road. We could go on foot but we had already lost the car. If the driver was Oliver's boss, the operation had been well planned to evade us, and I had been right about someone smarter than Oliver being the leader.

"How could we have foreseen something like that?" Ray said. "Even their timing was right, so we couldn't possibly have caught the one in the BMW."

"We couldn't have, but you're right to question this, Ray. It shouldn't have happened, and the next time we'll be playing it differently. I gave in to our client and allowed him to limit my use of the police. It was to keep Gladstone's problem out of the press."

"But they can fire us if we refuse to comply with that."

"We should have had police on their radar, blocking off streets in the area. I'll have to just take the chance of being fired. If I don't we'll loose the case, for sure. That's even worse than getting fired."

It was good to have Ray's young, idealistic mind, to help balance our older contingent. If it had been up to Willie he wouldn't have worried about it, and I would have continued to embrace the same wrong strategy. I used the short arm of the agency, when we should have relied on the long arm of the law. When I had worked for Willie he had frequently made calls like this one, but it would usually work out. He had been my mentor and I still had a lot of respect for him. I refused to blame him, as it had been my call.

"I really thought we were going to do it," Ray said, as we got back in the car.

Wanting no more discussion, I refused to respond. I remained silent most of the way back to Austin. To suffer was what I needed to do. As we drove back by those decorated oil wells a siege of irrationality gripped me. I wanted to run out and rip off the jig saw cut outs and smash them. Ray tromped the floor pedal as if he knew what I was thinking and got us back to the Convention center in a hurry.

Oliver was back at the Convention Center as if nothing had happened, and had even parked in the same spot at the hotel across the street. Ray dropped me where my car was parked, at the Center by the coffee shop. He went back to look for Denison. I checked my cell phone and found it turned on as it should have been. I didn't want to call Willie and take the chance he would be at a touchy spot when it would go off. He would call me as soon as he had anything. I had to be patient while it still burned inside of me. Off center, and needing something, I got coffee to go and sat in the car to do my sulking.

The cell rang so soon that I jumped in surprise and grabbed it up.

"Got some interesting news," Willie said. "Oliver has a new prospect, so how'd it turn out with you guys?'

"We lost whoever it was in back of the house and didn't even get a license number."

"You can't win-m all."

"No, but this would have been a big win and for the next one I'm calling for more help."

"That'd be OK, it's your call."

My call alright, and it started up again inside of me. I didn't let on, though.

"Have Ray take the new one. I'll keep up with Cal Denison and talk to you later."

Something was not working now between Denison, Oliver and the third party. It seemed likely the price was pretty high and Denison had to think about it, or consult with others in his company. He would probably want to do both.

The next thing I did was get on the phone with Carl Watson and as usual he was on a sour note. I could hardly blame him though, regarding the latest developments. At least he had to agree with me about needing more assistance from the police.

"We need to get back to using my men to turn up some new leads, and make things happen. You could use the Crystalville police to help us with some of the routine surveillance."

"How much surveillance are we talking about?"

"I have two men trying to watch three at the Convention Center. Can you get us two plain clothes cops and a roving radio car, ready to call in others to put up road blocks, when Oliver comes up with another prospect?"

"What are we doing here helping Oliver to sell stolen merchandise?" Carl said.

"You know better than that, Carl." I could see the big man was still miffed.

"Maybe I do, but we have to be careful bringing new people into this"

"The first thing I'd do would be to have a briefing with the new detectives, so they would get to know us, including our short term and long term goals on the case.

"I briefed the CPD about the secrecy thing they're trying to keep going at Gladstone, but you better go over it again," Carl said. "They've

been doing well with it so far, and you don't drop the ball, Brandon."
I said I wouldn't.

It could help a great deal to have the CPD in the know about what had happened at Gladstone. I'd continue to stress the secrecy about what we were doing and said so to Watson. He agreed to call out the help I had asked for from the CPD, and then call me back when he had the answer. I told him where I was at the coffee shop, and that they could meet my men right there, and by then we would have a private room in the Center or in the hotel complex for the briefing.

Dropping the plastic cups from the car into a bright colored container by the door of the coffee shop, I trotted over to the next building where I had noticed the administrative offices were located. I filled out a simple one page request for the private room. I included Address and phone numbers for the people at Gladstone and Carl Watson to represent whoever we were going to get from CPD. In a few minutes I left there with a printed card with my name on it, and the assigned room number, which was B1, in the Dowager Wing of the Hotel Complex.

On the way back to my car I checked the cell phone to be sure it was on and it was. I drove across the street to the multi-storied Hotel Complex, where most of those attending the Global Technology Conference were staying. I picked up a printed copy of the lay-out showing the locations of all the Main Ballrooms and subsidiaries. It was starting to warm up outside and the air conditioning in the main lobby was on. Since I needed the exercise, I didn't mind the hike to the other side of the building, to the Dowager room on the ground floor. To the right of the entrance double doors was a sign on the wall saying B1, by a short stairway going down. The small basement room would be perfect for our needs. There was a conference table and more than enough chairs. It was away from the main areas of traffic, so you couldn't ask for any more in the way of privacy. I searched my pockets and found some of the information I had saved about the conference, UT at Austin and about some of the Central Texas Companies getting the spotlight. I thought I had more in the car.

It should be easy to make up a printed packet of information for each of the CPD officers in our group, only about five or six packets including Ray and Willie. I thought about it as I walked up the stairs from B1 and back to the car. I did have more printed information that would be suitable in the car. I had been shown the materials, scanners, computers and other things that I could use at the administrative offices, where I had been a few minutes before, so I drove back there and parked in about the same spot. As I got out of the car with the printed information I intended to use, the cell went off again and I answered it.

"I got three plainclothes people with an unmarked radio-car from CPD." Carl said. "Where do I send them, Brandon?"

I told him the location in the Hotel Complex, where I had reserved the private room and he said they would be on their way. As soon as Carl hung up I called Ray's number and asked him where they were now.

"Willie is stuck in one of those presentation rooms where he's watching Oliver and the new guy. I'm outside of another room, where Denison is making a presentation."

"We have to leave Willie alone," I said. "Denison is in charge of one of the presentations, so he can't leave there for a while. We can call new guy Mr. X"

"The program sheet says it will be all day. He has a lot of computers and other equipment set up in there and he's providing hands on instruction," Ray said. "He's got several others from his company working in little groups. I don't know enough about computers to fit in any of those groups so I've just been staying outside"

I gave instructions to Ray about the Dowager room and B1 and told him to go there and be in charge till I could get there. If the others got there before I did he would show them his ID, and tell them why we were here at the convention. It didn't take long to run off six copies of my printed information at the materials center, get back in my car and drive back across the street to our private room at the Hotel Complex.

It was hard for me to believe that they had all arrived and were quietly waiting for me, when I walked through the door at B1. There was a chalk board and tray with chalk along the far wall, parallel to the table and chairs they were sitting in. I picked up a piece of chalk, wrote my name on the chalkboard, and handed out my information packets. I thought there would be plenty of time to show them my identification, and get friendly after I went over some of the things they needed to know.

"A few days ago Bob Clauson was abducted and his laptop stolen," I said. "This is something I can personally attest to, although we are being secretive about it as much as possible."

I held up the first page of their handout and pointed to the picture of a professor of electrical and computer engineering at UT in Austin. I related the information I had gleaned from an article about the inception of the Southwest Academy for Nanoelectronics. This research institute was receiving over a million dollars of start up money from the Nano Electronics Research Corporation. The state was also contributing about a million for start up. The professor at UT held a semiconductor wafer that was a sphere about half the size of his head. I mentioned what was said in their handout, about how the wafer contained millions or even billions of micro-and nanoelectronic devices.

"The hard drive in Bob's laptop had a special wafer like that in it, to make the jump from gigabyte technology to yottabyte technology," I said. "You will hear some people, who are working scientists in the field make jokes about the nanobyte, saying that it means no information. There are several bits of information in one byte. It is a play on words, referring to the inadequacy of the computer code. When the code was invented the bit was considered to be the smallest unit of information. It's merely relative, but the joke is the in thing now. A nanobyte, if there was such a thing, would be one billionth of a byte. There is that much information in what they call the super-laptop at Gladstone Electronics, and the age of small, fast computers is here."

Money has always been a great motivator whether we like it or not. It's connotations about good and evil have always been there. I returned to my discussion of how much research money was being given to UT at Austin, for a total of thirty million dollars. Ten million was coming from the State's Emergency Technology Fund. The Regents of the UT System were giving ten million. And the various semiconductor companies were giving ten million. I said that thirty million dollars in research money sounded like Nano Electronics was going somewhere. I paused for a few moments and did a visual appraisal of my small audience. They seemed attentive and intelligent. The three plain-clothes detectives from CPD consisted of a woman and two men. The woman had attained the rank of sergeant and commanded the radio car. That radio car would be able to call in the entire police network. I felt good about it and left my chalkboard to join the group.

Ray had already introduced himself. He introduced me to the detectives from Crystalville, who had been assigned to our case. We just chatted for a while and drank some of the ice water provided by the hotel. They were going to check into private rooms in the hotel and be working with Ray and Willie doing general surveillance. The names were easy to remember for the men, a Charlie Sloan who was sort of laid back, and a Joe Allen who was the serious, unassuming type. The woman's name was Christiane Amani and she said she went by Chris. She had the old-world look of a gypsy with heavy black eyebrows and long black hair she kept close cropped around her face and manageable. I would have liked to spend more time getting to know them better, but this wasn't that kind of a case. We were always pressed for time, and I was having an uneasy feeling about Calvin Denison. I had to find out more about him.

It should be easy enough to learn about Calvin Denison by just visiting him at the Comp-ease Computers Exhibition. Since we had been together through the messy murder of Tom Oliver I had given him some free advice about how he should be prepared for his police interview with the APD and for that matter if necessary with their little brother the CPD. Cal was receptive to my short counseling session with him and I had heard that he had done well and was not being considered as a suspect who had anything to do with the death of Tom Oliver or any illegal transaction regarding any kind of computer.

CHAPTER 10

It seemed as if I was destined to be moving back and forth between the Convention Center and the Hotel Complex. I went back to the Center and picked up a program showing the list of exhibitors and where they were located. Comp-ease was on the ground floor at room 105. Ray was standing in the corridor at a cautious distance of about fifty feet from the open door of Cal Denison's exhibit room.

I asked him if he'd be surprised to find out that Oliver had actually set up an appointment for the new guy.

"Yeah, guess I would," he answered. "You think it's a sure deal with Denison and the new guy thing is just a diversion?"

"We have to accept it as a distinct possibility, and it may be the most likely thing. I'm going to have Joe and Charlie working with you twenty four seven. You want to go set it up with them now, while I'm here to watch Dennison?"

As soon as Ray was gone I walked casually on past the door and glanced inside. One of Denison's men came through the door at about the same time. He was clad in a clean white lab coat with Comp-ease stitched over the pocket. All of them, but Denison wore those white lab coats. I thought they were computer salesmen and if someone acted particularly interested in any of their products they would send them to Denison to work out the particulars and maybe even sign on the dotted line.

It occurred to me that Gladstone might have an exhibit set up here. It might be something like Comp-ease was doing, but I hadn't the slightest idea. When I checked the program, I found them listed under the presentations with Dr. Von Mueller as the main speaker. The topic was Nano Electronics and it was scheduled for today at one PM at the main auditorium. It was during the general assembly for everybody. I was impressed but didn't think I would be able to attend. We were stretched a bit thin on our surveillance. I thought if Denison walked out the door there was no one else here to transfer him to. It came to me that if he did leave the room I should pretend to be talking on the cell phone, so it wouldn't be too obvious that I was watching him.

I decided to make a more permanent fix to our problem and called Ray. He answered on the second ring. "Are you with them now?" I asked. He said yes, and wanted to know if there was a problem.

"It's not too likely Denison would leave his room," I said. "But the unexpected always happens. One of his men just took a break and he might do it too. While on break he might think of something else he should be doing, and since he's the boss they'd cover for him."

"Oh, I get it, you want to keep two near Denison's exhibit so we can transfer and he won't get on to us."

He had the picture now, so I explained how I had planned on keeping Chris free to access her radio car, but she was going to have to do a shift at the Comp-ease exhibit too. He said the CPD people had rooms close together and Chris had invited them all to her room to organize things. She had wanted to be in on that surveillance from the start so it would work out fine. Ray and Chris seemed to be getting along well. I could see he had been taken with her. They were about the same age, with him maybe about five years older. I told him to tell her I wanted the two of them to start the first shift together now, so I could talk to her about the radio car. I ended the conversation on the cell but kept it out so I could stick it up to my ear and look straight ahead if Denison came charging out of there.

When Chris and Ray arrived, they saw how I was holding my cell phone at the ready position, and did the same with their own cells. I had Chris go with me a little farther on down the corridor, so we could go over things about the radio car. She had all of our cell numbers and we all had her number. A cell was a must in our business and it was especially important to have her car ready when we needed it. She had several years experience with using special technology to access the police network. I was confident in her ability to do the job, and I went back to check things out with Willie. There were so many of these high-techs with cells to ears, it would be good cover for us.

Someone had to relieve Willie once in a while. I had to be careful not to get so carried away with the other operation that I forgot about him. I found a side door and went outside into a small courtyard that was empty to make the call. He answered right away.

"Are you somewhere free enough to talk?" I asked.

"Yep, what's going on?"

"We got our help from the CPD here, now, and I'll give you the cell numbers."

I could here the rustling noise on the phone as he got paper and pen ready. I gave him the three new numbers and emphasized the importance of the one for Chris Amani and her unmarked police communications car. He laughed and said he would underline that one. He went on to explain where he was. It was my favorite coffee shop next to the main building of the Convention Center. Willie called Oliver's new prospect Mr. X, because he couldn't get a name for him. The two guys were in the coffee shop and Willie was watching the entrance door from a window inside the main building of the Center, where I had been using facilities of the administration office to prepare my hand outs. I told him I was nearby and would walk over there.

It couldn't have been more than two minutes and I was standing next to him looking through the window at the coffee shop. There apparently had been no change in the situation and they were still in there guzzling coffee.

"You weren't kidding about being close," Willie said. "It's good to know we have improved surveillance now, in case Oliver makes another appointment to show off the super laptop."

"That would be too easy to be true, but I wish. I think whoever is behind all this is a chess player, who likes to plan things in detail. All the moves are well planned in advance. I doubt that the same method will be used again, but we had to be ready in case I'm wrong."

"If this is an inside job, then that someone we're talking about may know all of our plans in advance, Willie said."

"I'm glad we have a radio car," I said. "It will widen our scope, and help us gain more information, even if we don't get to close the case with it." As I spoke my eyes were following someone who had just parked in the adjoining parking lot where my car was parked. It was an elderly man who was walking on the sidewalk towards us, and he looked familiar. As he came closer I could see it was Albert Hoffman, who had come to Austin Convention Center. An attractive woman was walking on the sidewalk coming from the Coffee Shop.

If they both kept going straight they would soon collide and now I could see it was Marsha Brooks going to meet Albert. There was no activity at all for a while then all of a sudden people were coming from everywhere. I glanced at the coffee shop entrance and saw Oliver and his friend standing outside lighting up cigarettes. "I have to go, and see some friends from Crystalville, Willy," I said. I took out my cell, held it up in front of him and carefully checked to be sure it was on. "If you get anything going with Oliver, and it's touchy just call me and I'll call Chris to get the radio car."

"OK, I'll do that," he said.

It was time for the main session at the auditorium. Everybody was getting ready to attend the one PM meeting, which would be led by the illustrious Dr. Von Mueller. I rushed on down the corridor like the "mad hatter" trying not to be late myself. Her melodious, sweet, southern-bell-voice sounded behind me. "Oh, Mr. Brandon," Marsha said. "Are you going to hear Dr. Von Mueller?"

My answer was an emphatic yes, and that I had looked forward to it. Well, that was a little white lie, but I was trying to be nice. God knows I still owed her that. She insisted that I go with them and I said I would be happy to. Whenever the corridor was wide enough she got between us, held one on each arm and steered us. Albert laughed when she did that, like he thought it was a funny joke but I knew that it wasn't. She just liked men that well and I thought it was something special.

When we got to the auditorium, she steered us to a place where there were three seats together. Just as we sat down my cell phone went off and I answered it. Willie said Oliver had left Mr. X and gone back to Denison. They left the coffee shop together and Willie was following them back across the street to the hotel. We closed our cells.

The introductory speaker took the podium and the auditorium quieted down some.

Marsha whispered, "We know you're doing your job and we're glad we got you."

As if to prove it, I ran out of the auditorium and called Chris. She was already standing by her radio car in the parking lot. Thank God for cells, I thought. Ray had given her the description of Oliver's car and the license number. She said that she could see it from her location and was ready to roll. I was running through the coffee shop's parking lot toward the traffic light on Caesar Chavez Street. I couldn't even see the Hotel Complex parking lot, yet, and was aware of what was going on there. We couldn't make it without our cells and the criminals had them too.

Slowing down to a jog to cross the street at the traffic light, I had the cell phone at my ear and was still talking to Chris. It was easier to go on foot from where I had been. It would have been much harder to drive my car at this hour. If I could get there in time I intended to hop in the car with Chris. As it turned out, I was there in plenty of time. She was standing by the car looking expectantly at me.

"He didn't show," she said.

"Yeah, sometimes it turns out that way." I didn't want to let on how I felt, and just smiled while tapping in Willie's number on the cell.

"They just sauntered into the bar for a drink," he said.

"You were closer, but I still don't know how you did it," I said. "I'm glad I got you, Willie." He laughed and signed off.

Chris was smiling now and I returned it. "He was a track star in high school," I said.

My breath was still coming in gasps as I told her Willie had followed them to the hotel cocktail room. "They were having a drink instead of attending the main meeting on nanotechnology being held at the auditorium. She lost her smile and was standing there akimbo, looking at her cell like she wanted to throw it. Turning around to retrace my steps back to the auditorium I had the same helpless feeling, and could only bite the bullet.

It was hard to imagine that any high-tech professional like Cal Denison would come to this conference and miss the main session. I thought he had something more important going on, like making a fortune on the super-laptop deal. If his company had that special laptop, this conference wouldn't mean a thing. The possession of the laptop could make him a new expert in the field of Nano Electronics, and he might even be the main speaker at the next conference. If he had been given a price for it and he thought he could come up with that amount, it would change things considerably. He might have been given a number to call, or a place to go at a certain time, to conduct his own business without Tom Oliver. But, why were they together now?

My best guess as to why those two were skipping out to the cocktail lounge was simply that they ran into each other at the coffee shop and decided they needed a drink. The new guy, we called Mr. X, had somehow fallen out of sync with Oliver, so when Oliver saw Denison he rekindled things with him. He must not have been sure if he was still in the deal if Denison did buy the laptop. Why not just hang around with Denison on the chance that if Denison was contacted directly he would learn about it? That must have been the reason they were back on the same page. I thought, if Denison had been given a way of contacting the seller when he was ready to buy, he would just wait till Oliver was out of the way before he made the contact. On the other hand, if he thought of Oliver as part of the deal, or just would feel more secure with Oliver there he might let him in on the contact information. I would have to wait and see how it unraveled.

They were into a short intermission at the auditorium when I returned. I had missed the entire presentation by John Von Mueller. The next speaker was Wallace Norton from the hole in the wall at UT. Someone had taken the seat I had with Marsha and Albert, so I sat in a seat in the back row and listened to the closer's presentation. I knew Norton was the closer, because he had said so while he joked around about how nobody could follow Von Mueller.

I thought, if nobody could follow Von Mueller then he sure couldn't. He was no comedian either, so I went back to thinking about Denison and Oliver. I pondered over their relationship, and came to the conclusion that something might happen soon with them. It turned out that Norton was brief. I left there as soon as he finished.

Heading down the sidewalk between the front of the auditorium and the administration building, I joined Willie back at the Hotel Complex.

CHAPTER 11

It was a surprise to see Norton getting into his car, which was parked in the lot at the rear of the auditorium. There must have been a quick exit from the stage that he had used. He got inside of his car and I jogged right on past him to the traffic light on Caesar Chavez Street. It was green, and as I started across I remembered something about Wallace Norton's car. It was a blue BMW. There might be thousands of them around the Austin area. It was no proof that he was the mystery driver who drove away from the scene at Lulling in a blue BMW. But it was something to think about.

There was a small inside shopping center at the first corridor to the left as you came in the front lobby of the Hotel Complex. The cocktail lounge was located in about the middle, and I could see Willie standing outside of a plate glass entrance door at the other end. To avoid being seen by Oliver as I walked by the front of the small bar, I walked back outside and around the front of the building to the east side entrance where Willie was stationed. He saw me and stepped away from the door so we could talk more private.

"We just got a break, Dan," he said "There's a back room to the bar just for the help. I went in that way and told them I was doing police surveillance and needed to get close to those two in the bar. Oliver's cell phone rang and he got pretty excited about the conversation. They're supposed to wait twenty minutes and then go to Oliver's room where there's a sliding patio door. They're supposed to open the curtain, unlock the door and wait for someone to come. In ten more minutes they'll leave."

"They could get another call telling them to go somewhere else," I said. "So you wait in the back room and listen while I get the others set up around Oliver's room."

In a few minutes I got the others out of the parking lot and had them stationed on the outside of Oliver's room far enough away and yet close enough to be able to apprehend the person. I didn't like the set up but we had to make the most of it.

I could see the motion of the long curtain sliding open, and hear the sound of the sliding patio door opening. There was a nice breeze, and they decided to enjoy it. They pulled a couple of light-weight chairs through the open door and sat down on the concrete patio to wait. They had a drink in their hand while they waited and moved their chairs out for maximum freedom and comfort.

It sounded like three shots and I thought I saw where it came from as Oliver went down. There was a wooded area about one hundred yards in front of the patio. Denison was calling nine-one-one, as the rest of us ran to the woods. Oliver had to be dead.

The shots had to have been fired from a deer rifle with scope. About twenty five yards into the woods we found the three shell casings. I picked them up one at a time by placing the point of a pencil into the space where the bullet had been before it was ejected through the barrel. I dropped each one of them into an evidence envelope and gave the envelope to Sergeant Chris Amani, who delivered it to the APD lab. From that vantage point we could see a parking lot on the far side of the woods, where the shooter must have parked the escape vehicle. There were no tire marks on the concrete parking lot, and no witnesses to the crime for us to question.

The small parking lot ran along the side of a service station. It was used to park customer cars. It was a full service station and some of the cars were being repaired. Some of the owners had gone home. We questioned everybody that was there and none of them had seen the shooter. One customer had been sitting outside in front of the station. He said that he might have heard the shots, but at the time he thought it was a car backfiring. None of them had seen a suspicious character with a deer rifle slink into a vehicle and drive away. And none remembered seeing a blue BMW parked there. We took names, addresses and phone numbers, so we could follow up on them later and went back to the hotel patio.

The Austin police ambulance had arrived and the body of Tom Oliver was being taken away with two slugs in the chest and one to the head. Chris Amani followed the ambulance to fill out the police report and deliver the evidence. We would all be available to the APD, but I was especially glad to have three of Crystalville's finest here on the scene. It was their case from the beginning, and I welcomed their involvement. I thought it would also be a plus to have all of the facilities of the Austin Police Department working with us. It was an end run to be sending Sergeant Chris Amani to meet with the APD, but it was also the right thing to do. She could answer all of their questions and be welcomed on board. They would probably know as much about me as they would want to know. They could call me to their office and I would go.

Calvin Denison was very pale and so shaky on his feet that we had to help him to sit down. Charlie and Joe supported him from either side while Ray got the chair under him and placed a drink in his hand. Willie pulled the other chair over in front of Denison, knowing I wanted to seize this moment to learn all I could from him.

"We were trying to protect you, Mr. Denison," Willie said. "You could see we worked together as a team, and our leader would like to talk to you a few minutes."

"Yes, I thought you must be the police," Denison said. "Thank you for that, or I could be dead too."

They all pulled away from him and seated me in the counselor's chair.

"I'm sorry we weren't able to save your friend," I offered.

"Well, he was really more of a business associate and not even a trusted one."

"I wondered about your relationship, as we were aware you were dealing with criminals." I paused for emphasis on the criminal aspect of his involvement.

"At first I was not aware of it but I had begun to wonder, when Oliver took on a threatening attitude toward me. The business venture I was considering was perfectly legal . . ."

"We know there was something very illegal about the proposition they made to you." I hesitated long enough to notice he was sweating.

"What proposition?"

"I'm saying a proposal was made and you're involved in it."

"It must seem that way, since apparently Oliver was killed because he knew too much. He knew about the illegality, I didn't. And now that I think about it, I don't think that I was a target."

"No, not this time but maybe next time. If you're thinking you must know you're on shaky ground. You're doing what anyone in your spot's got to do, and that's to get the story straight. The police have to think it's believable to buy it."

"What do you mean? I thought you were the police."

"Yes we are, but only the small town police from Crystalville," I said. "We're finished with you, but you'll soon be talking to the APD, our big city brothers. This is going to become a big case. You should get a lawyer. They can afford lawyers and several special investigators to listen to your story and poke a few holes in it."

"I'm not making up a story," he said.

"You may think its all true, but you can embellish the facts and not know it."

His hand was shaking as he accepted the business card I offered him. If they asked me I could verify what he told me, and even add some of my own views. It could only help him with the APD if he told them the same thing he told me. I was betting that he would tell it close enough to be believable.

That dangerous dusk time had come and gone without me even noticing. When too many things happen at once we tend to put everything on hold till we can get a handle on it. The subconscious mind is always ready to do its part by giving us some of those convenient short memory losses, and we feel better than we really are. In reality the process has left us physically drained, like driving half the way to Dallas in a fog then waking up to see how far you've gone. You can't believe it.

I left my car parked where it was at the Convention Center, walked the few blocks over to Claire's apartment and was standing in a daze looking down at her scrawled note on the dining room table. The firm had sent her to Houston on a new case. She would give me a call when she could, and I was to help myself to some of the things she described in the refrigerator. It all was good stuff and she had gone to more trouble than she should have. I just didn't want to think about it. When you have just witnessed a dead body being carted off like a side of beef, it does something negative to your digestive system. Oliver hadn't been on my favorite list, but he had been a human being and the dead condition was one of the most pathetic human conditions I could think of.

The apartment was so quiet that every small sound seemed increased to at least the tenth power. I came close to turning around and going back to work. I was aware that I had put in more hours already than I was capable of, considering the weakened condition I had been in since I received the sound beating about my head and body. What I did was something I don't do often enough—and that's the right thing. I drank a half a glass of sangria and went to bed.

When I wakened I had come out of a depressing bloody dream like the day I had just spent, and I couldn't remember much about the details of the dream. I was thankful for that, and felt rested enough to get back into things. From the menu in my mind I selected the Continental Breakfast. Claire had filled a large carafe of coffee before her departure, and I poured myself a cup while I waited for the roll to warm up in the microwave. I took a quick shower and dressed using the extra set of clean clothes I kept here. It was time for me to give myself the famous half-shave with my electric razor. It would probably do more for me than it did for Brad Pitt.

I remembered Marsha had brought me my old Norelco when I was supposedly recuperating in Crystalville. Claire had given me a new one that I kept at her place, but I had been ridden with guilt over having Marsha bring me things I needed from my office downtown. I didn't want to have to explain things about Marsha to Claire. I had done nothing wrong. It was just that I was working too hard at being right. I wondered what Marsha thought, when I had materialized at the Main Session of the Convention. She must have thought I was so devoted to Gladstone and the good Dr. Von Mueller that I got up from sick-bed to attend his speech entitled Nanoelectronic Devices. I wondered if one of them had jumped in a car containing a nice compact deer rifle and sped around the corner to the service station parking lot, and put an end to their relationship with Tom Oliver. Norton had to be number one on my list of suspects, at this time.

If Wallace Norton was behind the computer theft and the abduction of Bob Clausen it would greatly simplify things for me. But I started thinking what if games about Norton having an accomplice at Gladstone. It would have been almost impossible for him to have orchestrated all that business while he was working on the research team at UT Austin. His ownership of a dark-blue BMW did not constitute any sort of involvement in a crime. He wasn't any great shakes but he was all I had, and I still had to do a great deal of investigating to solve this case. I couldn't say I wasn't getting enough help now. It was too late to be able to walk away from it.

Thanks to better communications provided by computers and cells the CPD would soon have something on those shell casings. I thought of the similar shooting done on the Bastrop Highway with Campos the security guard as the victim. I called Willie.

"I need more details about the way George Campos was shot," I said.

"I'm not surprised to hear it, Dan, and I already talked with the APD lab. They said the bullets were from a high velocity weapon, probably a rifle. The man at the lab said they dug one of the slugs out of Oliver, compared it to one from Campos and they matched. He said they were working to quell the rumor from getting started that we had a deer-hunter serial killer on our hands."

"Did you tell him to check with Chris Amani for a possible motive?"

"Yep, sure did. Oh by the way, the CPD called back Charlie and Joe."

"They'll probably keep Chris too."

"You got it figured right, buddy."

I told him to start it out by the two of them, Ray and Willie, to follow the blue BMW they would find parked at the hole in the wall place on the UT campus. I described the location with the red striped saw horses with the blinking, battery driven safety lights and told him how to get there. It was another one of those signs of the times, when car dealers sold more BMWs as a result of nanotechnology. New light weight materials were used in the bodies of these full sized cars and they ran on non-petrol fuels or electricity, which was finally made cheaper than the old air-polluting petrol.

CHAPTER 12

My car was parked by the Liberty Café in one of those parking spaces, which mingled with the spaces for the adjoining Convention Center. There always were enough spaces and you didn't get ticketed or towed. I already had my continental breakfast or I would have gone in for a vegan chocolate-chip cookie. Those things are said to have enough energy to see you through a revolution. It would have been a very revolting development if I had parked the car somewhere else, and had to go looking for it. I wouldn't have been surprised to find that I had left it on the other side of Caesar Chavez at the Hotel Complex.

The cell sounded off as I got in the car and I answered it. Nelda was calling from the UT lab. She said she had just received a call and was told to take the current version of her research-laptop to a place in Luling. The caller said she should do it right away, or something might happen to Bob Clauson. If she told anybody about it something would happen to her too. "You did good to call me, Nelda," I said. "You have to refuse to be intimidated while you play along."

"I'm supposed to go to one of those permanent flea markets. I get on the main street as I go into town, it's a tree name and I take it till I come to James Street. The flea market is right at that corner. It sounded like a man. He said he'd be looking for me."

"I'm near highway 183 now, Nelda," I said. "I'm going to be in Luling before you. I've been to that flea market and know the layout. Running north and south are Avenues with the tree names and east and west are streets. There's no Main Street. That would be Magnolia Avenue. It goes through the center of the town."

"There could be several men waiting for us, Dan," she said.

"Just let me worry about that. I'll have back up, but you won't see them. If you don't see me just keep going around the block, but do not get out of that car under any circumstance until you see me."

She said she would do as I asked, although she was talking in a high pitched voice and was very nervous. As soon as I signed off with her I called the Luling Police Department and explained the situation. The duty officer was a Captain Torres. He made a quick call to CPD to verify my case with them and called me right back. He said he would have men stationed around at all of the exits ready to put up road blocks. When I was in place I'd call him back and he would tell his men to stop everyone going out of town.

It seemed to me that we were prepared to catch somebody with our makeshift trap, but I didn't understand what it was all about. As far as I could tell, there had been nothing going on in the way of research at the UT lab, so if someone had the real thing why check out small potatoes. And it would present a delicate situation for the criminals, increasing the risk of getting caught. I thought about how my own risk was greater as I sped along on 183, in the left lane with my foot to the floor while I called back Nelda.

"Yeah, that you, Dan,"

"Remember our plan?"

"I keep going around the block till I see you walking there."

"Right, Nelda," I said. "What would they want with your laptop when they have the super-laptop?"

"They probably have a buyer who is aware of how we moved improvements on the module in my laptop over to the super-laptop, and the buyer wants to be certain it's up to date."

"That implies they know a great deal about the research, and I thought it sounded suspicious."

"When you were at the lab the other day, I mentioned the possibility that this might happen. Do you remember that?"

I said I did, and asked her how someone would know her cell phone number, to call her at the lab. She told me a lot of people had her business card and now she could see it should have been a security precaution. She should not have been handing them out so freely. It had been left up to her to handle her own business cards, and so much for hindsight.

I was coming into Luling and had to slow down and get off the high-traffic freeway.

"I'm coming to the cut off for your Street," I said. "I've signaled the local police to set up road blocks for anyone leaving town. I'll park and be there for you."

The week day mornings didn't seem very busy on the main drag in Luling. I found a parallel parking space about two blocks before I reached the intersection that was my destination. I got out and started walking toward James Street acutely aware of my surroundings. There were department stores, pharmacies, groceries, and fast food restaurants. I tried to be relaxed and casual, glancing to the left and right as if reading the advertisements. I would occasionally look inside of the establishments as a typical shopper might do, and paused while I checked out the merchandise.

I was reading the menu displayed inside of a pizza place and it was making me hungry, when I heard a car drive up and park on the curb. It was Nelda in a red BMW with the windows down. She cut the engine and just sat there looking around. Several teen-agers went by on the sidewalk between us. They were talking and laughing about something only they knew about. There was no one else near us, but I ignored the young lady in the flashy red car and looked in the next window. I could hear the car start up again and the windows closed. She had forgotten to close the windows and had to have the engine running to use the electric window closers, and then she locked it up and got out.

She started walking up-the main tree street toward James Street where the flea market was located. I let her get a little start on me and then continued on that way myself. I noticed that the BMW seemed to be a popular automobile for the research staff at the UT lab, and that she carried the laptop in a black trash bag. I thought the latter was a good idea, because a street thief would be less likely to grab it. A thief is more cautious than the average citizen.

Nelda was moving slowly up the slight incline we had encountered. It allowed me the time I needed to scan the area as she came to it and move along at about the same pace. So far our plan was working well, and I had no reason to think it wouldn't continue to work as she came to the intersection and crossed over to the entrance of the indoor flea market. This was where I expected something to happen, a car to appear there with a driver and at least one other to grab her and the black plastic package. They would know what was in it, and they had to know Nelda well enough to recognize her. I didn't expect her to be accosted inside the building, but there was no vehicle of any kind in the street. I crossed the intersection as she came to the door. A man came through the door reading a shopping newspaper and collided with Nelda. He seemed high on something and kept laughing and talking to her about the way she hung on to the black package like a football, as she was going down with it. He had caught her before she did any damage to herself. He wore a grimy cap, worn out jeans, an orange shirt and dirty sneakers. His antics seemed innocent enough, but I was ready to go after him if he took off with our black package or bothered with Nelda in some other way.

"I guess I should watch where I'm going," he said.

"Well, while you're reading something you should stand still," Nelda said.

"I'll remember to do that and I'm very sorry. Are you Ok?"

"Yes, I'm fine."

The man continued on his way outside and Nelda went inside the large building. I counted three seconds and went through the door after her. She walked through all of the flea market booths and exhibits. I moved along staying about twenty or thirty feet away from her while I looked around for anyone who seemed interested in Nelda and her black bag. She made another complete trip around the place to satisfy herself that she had been there long enough for someone to have contacted her.

When Nelda finally left the flea market, I watched her through the glass from the inside of the large entrance door, to be sure that nobody was waiting outside and would fall in after her. It didn't happen. There was no welcoming party at her car either. She was almost all the way down the slope on Main Street before I went outside after her.

She was sitting inside her car with the window up when I walked by.

"What do we do now?" She said through the window.

"Wait till I drive by, then follow me out." I said. "I'll talk to you on your cell." I kept walking with my eyes straight ahead in case we were being watched.

There wasn't anyone waiting for me at my car, and no notes on the windshield. I got inside and drove past Nelda's car as I had said I would, and she drove her red BMW behind me to the outskirts of the town where I was pulled over by a uniformed police officer. She got out of her car and came over to talk to the officer also. I explained about her black box and how we were responsible for his road block operation. I was going to follow her back to the job at the UT campus and have Willie meet me there. He would hang around for at least the rest of the day and see her safely to her home.

When the Luling road block ended and our business with them was all cleared up we started back to UT Austin, in our separate cars. In only a few moments I got a call from her saying she was just following my procedure to talk and she had something.

"It's about that guy that bumped into me," she said. "When he was jostling me around I think he was checking to see if I carried a gun, because he said in a low voice that he was Tom Cobb with the FBI, and if I needed help I could ask for him at the Federal Building."

"I'm not surprised to learn about it, Nelda. Their agents are great actors, and he pulled it off perfectly."

"Then you're not mad at me for not yelling FBI right there on the spot."

"Certainly not, you did the right thing," I said. "I would not have fooled around with him then. He didn't want to talk to me. I'd have done the same as I'm going to do now. I'll go over to the FEDS and talk to him on their terms, unless you would rather do it yourself."

"No, I wouldn't. I've been a nervous wreck during all of this intrigue and I'm happy to have you handle it."

I told her that I would tell the FBI she was kept busy in the lab and she had asked me to talk to them. It was a sure thing that they were into this case. I didn't think Tom Cobb was his real name and as far as handling him, that was a misnomer. You didn't handle the FBI, they handled you. All I would do would be to hold my own and try to learn something from them that I didn't already know. I didn't tell Nelda that.

Willie was already standing by his car talking to the security guard, when we both returned to the hole in the wall place. He said there had been no blue BMW for him to follow. Nelda explained things to the guard and he slapped some kind of temporary parking sticker on Willie's windshield. That allowed him to accompany her inside and wait till the end of her shift. Willie had already called Ray and filled him in on the situation. They would be watching her place during the night, and I would be joining them the next day. That left me free to get Nelda off the hook with the FBI.

CHAPTER 13

Driving only a few blocks from the UT campus, I went downtown to the federal building, and parked near the high-rise building. I put money in the meter, leaving the engine running so I could go back to sit a while with the air conditioner on.

This was not something I wanted to do for myself. I'm sure I would have been better off to have steered clear of the FEDS, but I had to do it for my friend Nelda. She had let me know at the outset of this case that she would do anything to help me solve it, and remove Bob Clauson from whatever kind of bondage he was in. She had no way of knowing that she would receive a phone call that put such a great demand on her nervous system. We are all different regarding our reactions to stress. I could see it was just too much for her and I had to give her some slack.

The inside of the old building was huge and intimidating enough, without having to ask to see a man by the name of Tom Cobb. I didn't expect there to be any such person. I thought the name was being used as a code. If Nelda had come over here asking for Tom Cobb they would have said he was not available, and had someone else to work with her who had been specially trained to meet all of the exigencies of this case.

Keeping all of this in mind I asked for agent Tom Cobb, and in only a few moments he came to the outer office with the same devilish smile and wearing the same threadbare clothes that he had worn at the flea market.

"Come on back to my office, Mr. Brandon," he said. "You may sit in the comfy red-leather-chair in front of the desk."

"I bet you call it the hot seat," I said, as I sat down. He sat down behind his desk.

"No, the bureau wouldn't allow us to resort to cheap-shot puns but they choose the furniture. I hope Nelda Whittington is well."

"She's not so well, actually, so I'm here in her place."

"Oh, not well, sorry to hear that."

"Nelda does fine as a research engineer, but she can't stand up under the pressure of the emergency situation she's in."

"Yes, I understand, so she has asked you to touch down with us in her place."

"That's right.

"It will only be a short visit now and then. We've observed her enough to know she has a bad case of the jitters, and we won't place the extra burden on her. We'll accept you in her stead."

"So, what is expected of me?"

"The next time she gets a call like the one she got today, requesting her to take that research laptop outside for a stroll she will call you as she did today. You will call us immediately, since you are her friend and you want to give her the maximum protection that's provided by the FBI."

I said that I did want to provide her that protection and I agreed to call them the next time it happened. Agent Cobb left me thinking about what he meant, when he constantly referred to himself in the plural, as we. He said we, because he wanted to impress me with the idea of the multi-faceted powers of the bureau. I was well aware of the truth in his supposition. The one who called Nelda and threatened her into taking the laptop to Luling must have seen Cobb, smelled FBI and fled from Luling.

The woman who had met me at the door was gone and I left the building. As soon as I said I would inform him if Nelda had to take out her laptop again, Tom Cobb had left me to see myself out, and had gone down the long hall to another office. I walked back to my car and found there was still a lot of time left on the parking meter. I climbed inside and rolled down the front windows to enjoy the cool breeze and think things over.

The feeling that I had imagined the entire episode came over me, but Cobb was real and would come back into my life at any time that would suit him. It was for me not to worry, and I didn't. He had plenty of time to check up on me before I arrived, so it was no surprise that he padded up the chance to interrogate me. The FBI was one of those things we called a constant. Once you had worked with FBI you got familiar with their general mode of operation, and it was just a matter of working within their necessary constraints. You could plug them in or they could plug you in and the results would be quite predictable. It would be just as he had said. The visits would be short and only now and then. I was sure of it.

My cell had been turned on and I wasn't getting any calls. When the weather cooperated, and allowed me to use my vehicle as my office, the glove compartment was my filing cabinet. I had saved some of the small cardboard boxes containing checks that were mailed from the bank, and put them in the glove compartment. One of them I used for current business cards I had picked up from people on this present case. I was getting a lot of cards already. Cobb had seen me pick up his card from the front of his desk, so he didn't offer me one. His trained eye didn't miss anything. I dropped it in the box and took out Chris Amani's card. I tapped out her cell number, and she picked up right away. I told her I was concerned about the death toll on this case and asked her to do some talking around with her contacts from the APD. She said that she already had.

"When they found the bullets that killed Campos and Oliver came from the same rifle, Carl Watson wanted to know about the other member of their gang."

"Junior Gilbert," I said.

"Yes, that's the one, and so they searched around the wooded area near the spot on the Bastrop Highway where the body of George Campos was found. There's a bike trail not far from the highway and by a wrecked motorcycle they found the body of Junior Gilbert. He had been shot in the same way as the other two, and they got a match again from the bullets. His body hadn't been found sooner, because he had rolled over the cycle. He and his cycle were found in the woods when they searched the area this morning."

"Did they have any ideas about why Campos and Gilbert were around that end of town when they were shot?"

"They think they were both joy-riding on the motorcycle using that trail, when they were shot off of it. Gilbert died there deeper in the woods with his cycle, but Campos had only been hit once and he was running from the trail back to the highway when he was hit two more times. He had been hunted down on foot and the last two rounds were at close range."

"It was an experienced marksman," I said.

She said that the consensus of the thinking from the forensic team had included a ballistics expert, who had done a lot of deer hunting in all sorts of terrains. According to the expert the murder weapon would be the ideal deer rifle, and could accommodate every member of the family. It would be light weight and probably a Remington bolt action with a 22" barrel. It was chambered for the 7mm 08 Remington rifle. And most of these deductions were made by observing the evidence I had sent them. The shell case was a 308 that had been necked down to fit a 7mm bullet. The charge in the rifle produced mild recoil, and wouldn't kick you on your rear end before you could get off a second shot.

I had to agree that it was that kind of a rifle, since the murderer had been able to fire three rounds in rapid succession, with such precision. It would have been easy to toss in or out of a vehicle, and be light to handle in flat lands or the mountains. Good for hunting elk, deer, or humans, the ultimate in versatility.

Chris Amani continued to be a great asset on this case keeping me informed about the recent findings in Crystalville and the APD. Thanks to her close coordination between those two police departments I was aware that all three murders were related, and of more specifics regarding the type of deer rifle we were looking for. I confided in her about my recent encounter with the FBI and how it had come about because of Nelda's threatening cell phone call.

"Do you think the man who called her is also our deer hunter?" Chris asked.

"She thought it sounded like a man, but we don't know that for sure."

"For that matter, it could be the same person," she said. "We just don't know about the gender."

"Yes. It's likely that the murderer and the one who is trying to sell Gladstone's super-laptop are one and the same person."

"What was the FBI agent like?"

"He gave Nelda a break. If she gets any more calls to take out the laptop, she tells me and I call the FBI man."

"So, how do you think you'll like working with him?"

"At first I thought he was a jerk, but it was just part of his act. He was pretty straight with me and let me know subbing for Nelda would be light duty.

"Oh, Dan, it's time for our detective's meeting. We'll be talking about these same things. Carl Watkins will be here too. I have to go now"

I told her to enjoy her meeting and closed my cell. I thought about the one we had called Mr. X, the new guy that Tom Oliver had seemed to have lost interest in. For all I knew, maybe X had slipped right past us and had already contacted the one who had possession of that magical new laptop that every computer manufacturer in the world would like to get their hands on.

Driving a few miles farther south on Red River Drive, brought me back to Caesar Chavez Street and the Convention Center. I parked again in the adjoining lot at the Liberty Café, and started to stroll around looking for Mr. X. I had only seen him once with Oliver, lighting up cigarettes outside of the coffee shop, but I was sure I would know him when I saw him. The man was tall like Cal Denison and would weigh over two hundred pounds, about the size of Von Mueller and I thought he resembled Henry Kissinger. He had not dressed casually like Denison, but worn a suit like Von Mueller.

After I walked around for over an hour to no avail, I went to the Comp-ease exhibit to talk to Denison. I was glad to see he had survived his interview with the APD, and seemed more relaxed than he had been last time. He was seated in one of the portable chairs that had been set up to the side of the tables with electronic devices. He motioned to me and I joined him there. He threw up his hand and shook mine like a life long friend.

"Is there anything I can do for you, Mr. Brandon?"

"I'm looking for a man at this convention I thought you might know," I said. "He was having a cigarette outside of the coffee shop with Tom Oliver earlier today, before you and Oliver went over to the cocktail lounge and then all those other things happened."

"I do know him," Denison said. "He doesn't drink, or he might have gone over there with us. We had all been together in the coffee shop and he and Oliver both smoked. They grabbed a quick smoke and then James went to the main session to hear the CEO of Gladstone talk about the Nanobyte Electronic Devices. His name is James Lebeau, and he's my boss the CEO at Comp-ease. He got his PhD recently in electrical engineering and is working on a small yottabyte prototype laptop like the one missing from Gladstone."

"Is he here now?"

"No, he's gone back to the factory."

"He went back to Dallas?"

"Actually, he is at our new start-up division at Lulling."

It didn't take long for it to sink into my brain. I terminated our conversation in an instant, and left poor Cal with his mouth agape as I rushed out of there.

In spite of the usual traffic I made good time on the road to Lulling. The jig saw cut-outs were bobbing up and down in their usual celebration, like they were thumbing their noses at me. The new factory was on the other end of town where I had not been on my last trip here. The factory road was an extension of the main street. The building was constructed of gray stone and was nowhere near as big as Gladstone, but there was a black-top parking lot in back with quite a few employee vehicles. It had to be in operation.

There was a middle-aged woman sitting in the outer office in front of a computer screen with a cell in her ear. She looked up to motion me to a side-office, when I asked to see James Lebeau.

"Cal Denison referred me to you," I said.

"Referred you to me?" Lebeau asked.

"Yes, I'm Dan Brandon. I sort of helped Cal get through a rough time today."

"When I was there he did mention your name."

"Would you be turning out a new nano-electronic laptop?"

"As a matter of fact we are," he said. "We have been working on it for a long time. This is a special day. The first few are coming off the assembly line here right now. I haven't even got the sign up yet with our logo on it."

I had that don't know whether to laugh or cry feeling.

"Did Cal Denison tell you anything about what was going on around there?"

CHAPTER 14

The door to the administrative offices was standing open, when I got back to Gladstone, so I walked on into the main reception area. Marsha must have had the day off. Remkin stuck his head out of his office and motioned me back there.

"I'm holding the fort all by myself," he said. "How did you manage it?"

My head was hanging so low it felt like my chin would rub a hole in my chest.

"I guess you mean how I managed to get three people killed and your laptop passed around all over the state."

"We just got it back. Marsha and Von went over to the UT lab to get it. Von just called a few minutes ago to verify that he had checked it out and found it to be the real thing. They're on their way back with it right now."

"I'd like to take credit, but I know nothing about it."

"Nelda said she just got a phone call to go outside and pick it up. When she got out there, a man who was a street person gave it to her. He said he was given five dollars by a man to deliver it to her. She asked him to describe the man that paid him, and he ran away."

"Is Bob here too?" I asked

"No. I wish that he was part of the deal, but we still have to find him."

It was time for me to take a deep breath and give Remkin a complete report, which I did. I was surprised that he took it so well. He reminded me that we had made progress and that I was not giving myself enough credit. He was satisfied with my work and wanted me to continue on with the case.

"When you started working for us, we were missing both the inventor and his invention," he said. "I'm sorry to hear that another company is already producing the new innovation, but Von thinks we can still get back into it. Nelda is coming over here to help us out. He thinks with her help we might be able to start up next week, and of course when we get Bob back with us our chances will be even better, so get back out there and find him."

I said that I would and wasted no time about it, before he changed his mind.

I punched out Willie's number on my cell phone and informed him about the latest developments in this case. I told him about the new computer factory in Luling, the location and everything I knew or suspected about the entire operation. There didn't seem to be any reason to be guarding Nelda now that the laptop was back, so I said for him to take Ray and check into a local hotel in Lulling. They would shake a few bushes around there and keep me informed.

My desire was to remain in Crystalville and shake things up around there. I got back in my car and drove the short loop between Gladstone and my room at the Crystal Palace. I wanted to renew my relationship with Albert Hoffman for one thing. Now that Gladstone Electronics had shared the limelight with the other Central Texas sponsors of the Global Technology Conference, I had something to talk about to the local stockholders. If Albert Hoffman didn't own many shares of Gladstone, I would eat my sombrero in the middle of Caesar Chavez Street, after I announced my intention of doing so in advance so as to raise a crowd.

That twilight time had come and gone again without my notice. Lake Crystal sparkled with the reflections cast from several spotlights attached to the main hotel building, as I made my way up the lighted sidewalk to my room.

"I see you're back," Hoffman said. "I'm over here changing light bulbs."

I walked over to where he was standing at the back corner of the building. "Did you enjoy the Global Technology Conference?" I asked.

"Yes, I enjoyed it a great deal, and what was your take on it?"

"I was overwhelmed by all of that technology, especially the nanotechnology. Gladstone was well represented by Dr. Von Mueller at the convention."

"It was a shame that you were called out before he even started his presentation."

"I have a summary of it, and a lot of other literature provided by the exhibitors regarding nano-electronic devices."

"There is a lot of it here in my suite. I have been collecting information for some time. Would you like to see it?"

He took me into the front alcove and through a door behind the registration counter. Two large rooms had been combined with a small kitchen and living area. There were a lot of photos on the wall and framed clippings about Gladstone Electronics. Some of the larger eight by ten photos revealed Albert with his Gladstone friends, in full hunting regalia, ready to board their vehicles for a deer hunting trip. Another one showed them down at the Texas coast doing some deep sea fishing.

"The kind of nano-electronic research they've been doing will soon bring about huge advancements in our society. Your home PC will allow you to do a lot more than provide games, home security and research. Can you imagine PCs that will allow you to use new medical devices in the home? Medical devices have allowed people to bypass damaged nerve systems and move an arm or leg with the electric thought impulses of the brain. It's my understanding that the increased information and device technology will allow us to do that same sort of thing with our home computers. That would mean we could change a security code or a password just by thinking about it."

I told Hoffman I was aware of what big changes in our everyday life could be on the way, and that I was afraid some of it could be detrimental to our well being. Obesity was already a national problem. It seemed to me that as things got easier and easier we got less exercise. There was even a television commercial that extolled the virtue of hitting the "easy button."

There was no doubt in my mind that we were headed for such things in this country. The proof of it was self evident. You can see a planned sales campaign going on every day to sell us on an easy button of some kind. It's being led by our industrial leaders and politicians.

"Not to change the subject," he said. "I heard there was some sort of shooting accident at the Hotel Complex across the street from the Convention Center."

"Yes, I think a man was cleaning his gun," I said.

Hoffman raised his eyebrows. "I thought you would probably know all about it."

"I don't know all about anything in this case, and much of what I do know I don't understand. I'm not supposed to talk about it either."

"Yes, I can see how that could be," he said.

Hoffman explained all of his clippings and talked about the travelogue that was suggested by his collection of wall mounted photographs. I listened carefully and learned more about his personality. When I made the slightest comment, I could see his eyes glaze over. If things weren't related to something he'd said or was getting ready to say he lost interest. Now that he had finished his routine he was ready to get back to changing light bulbs or whatever needed doing around there. I said it had been fun, but I had to get back to work.

I stopped at the desk and found that I had no calls or messages. It was almost midnight and I realized that I had been putting in long hours, again. I went next door to my room, fell into bed and went to sleep.

It was about eight the next morning, before the sun reached high enough to finger me through a space at the bottom of the blind. I got up and ordered two eggs with toast and coffee, took a shower and put on a fresh sport shirt and a pair of kaki pants. I stepped outside to the sidewalk in front of the office, put coins in the machine and removed one of the morning editions of the Austin Statesman. I took it back to the room, leaned back in the recliner and read every page. There was no mention of someone having been shot and killed at the Hotel Complex of the Convention Center. The APD had successfully put it on the back burner.

The special treatment room was hard to beat as my third residence. It sounds strange to say it but I really didn't mind having three places in the Austin area. On another recent case I had done something to get the mob crossways with me and I found gunmen were set up to greet me at my downtown location, so I had to go somewhere else. I could have used the Crystal Palace, my special needs type hotel.

However, at the time, the recliner chair was so comfortable that I was going to have to get out of it. There were several computers around here, and I could take one of them in my room supposedly to work on this case. The next thing, I would be playing a few games of solitaire and ordering a stack of pancakes. It was the delicious taste of sin.

I got back in my car and headed south on LH-1 and drove the thirty minute trip to Austin. It could easily turn out to be longer, depending on traffic. Taking my usual route down Lamar put me at the hole in the wall place at the UT campus in a few more minutes. There was no car with a campus parking sticker parked there as it had been before. The space up against the building was vacant, so I parked there. The campus security guard looked at me with recognition, and turned around to go back to guarding the blinking yellow lights and campus congestion.

I rolled down the window and cut the engine but did not get out. I was receiving a revelation, a message from out of the blue. It said that I did not need the luxury of a special room in Crystalville, so why was it even offered to me. Nobody seemed to care what I did, and Marsha Brooks was sent to bring me whatever I needed. She had encouraged me to stay there in the room and rest. Albert Hoffman had been around the room watching me from the time I first checked in the Crystal Palace.

When I returned last night he was doing the maintenance man's job replacing light bulbs, and he had appeared out of the shadows to greet me. I had only his word that he was changing bulbs. He could have been just waiting for me. He might want to learn more about how the case was going.

Just maybe, I wasn't paranoid. They had been really watching me and trying to keep me out of the way, where I would do the least amount of damage. It was as if they had all agreed that I was a very bad accident, waiting to happen. Even Carl Watkins, who feared for his job had wanted me off the case.

Now, it was time for me to have a talk with Wallace Norton, but I realized that the lab room would be locked up. Since Nelda was leaving that job and moving over to Gladstone, the cell she had been using might remain there. I punched in the number and Norton answered.

"Brandon here," I said. "Can I come in there and talk to you for a little while?"

"I'm finishing up in the lab and getting ready to go through the tunnel on the opposite side of the lounge where we always meet. It's locked on your side and I had just finished locking up the other side. It's good you called me when you did. I'm on my way to teach a graduate seminar class, and when I get there I close the cell and put it in my desk drawer for an hour. If talk's all you want, it takes me about ten minutes to walk it, so go ahead and talk."

"What happens over at your lab, now that Nelda is at Gladstone?"

"We do nano-electronic research the same as before."

"It seems things have changed a great deal. Who takes her place?"

"I have a new graduate assistant who is moving into her place and Nelda has really graduated into a fine position at Gladstone. Oh, I think I know what you mean. They said her new job would remain secure, when Bob Clauson returns there. They'll retain her at the same rate of pay, whatever that would be."

"I'm thinking you're going to be hard to find, lucky I called when I did," I said. "You don't even park your BMW on that side anymore do you?"

"My God, man, changes are always going on. The graduate classroom I have started teaching is on that side of the campus and there is a parking lot near that location, and it is not in such a crowded area as the old place. That's all there is to that, nothing to do with our lab. The cell phone number in the lab isn't changing. If for some reason it did I would tell you."

The ten minutes he said it took to walk to his classroom was about up, so I cooled him down by telling him I didn't think I would bother him much, as long as I could reach him if I needed to. It was that cop thing. He said that he understood what it meant to be in my position.

CHAPTER 15

There had been a lot of my cases with many more miles of travel than I had logged in on this one, and I was very happy about that. But I didn't come out so well, when I compared what I had gained through the convenience of locations on this case, to all of the complications and unwanted changes. I would have been willing to have made several trips to the other end of the state if it would have allowed me to finish this case. What I did though was go back to Luling, where I already had two good men. I guess if I had ten more of them I'd have brought them along too. This was where something needed to happen.

Driving fast, but not fast enough to get a speeding ticket I drove to the south end of the city. The new factory was at the end going out of town. The Comp-ease sign had been erected in the middle of the front wall, centered a few inches above the entrance door. It read Comp-ease Computers-Laptop Division. I went through the door and just yelled out like I was in a loud barroom trying to be heard over the rest of the drunks.

"James Lebeau," I shouted. "I need to talk to you."

The petite, middle-aged lady smiled and looked up from the stack of print-offs she was working with on her desk in the outer office.

"He's gone out to get something they need for the assembly line," she said.

"That was what I was going to see him about," I said, lowering my voice.

"The foreman's here, in the back, if you want to talk to him." She motioned to a door standing open. "James told me Jeb Smith was in charge till he got back."

Thanking the lady I went through the door to the laptop factory. I felt guilty about the way I had charged into the office to see if I could stir up a couple of paid mobster type goons. I had just taken for granted they would be there.

A tall, intelligent-looking man met me, as I walked into the factory.

"You must be Jeb," I said. "I can't believe James didn't put some security in here with all this expensive technology."

"Don't tell him that," he said. He straightened up, looked quizzically over the top of his stylish black rimmed glasses and ran the fingers of his left hand through his blond hair. "James is the most sensible man I've ever known," he added. "He said for that kind of trouble to just call Captain Torres. That he would be all the security we'd need for now."

"I know of him as a matter of fact," I said. "He's a good man."

"Well I guess you know me, but James didn't tell me about you."

"I met the Comp-ease guys at the Global Technology Conference, Dan Brandon."

He extended his hand and we shook.

"Tell him I dropped by," I said. "I think he'll remember me."

He said he would tell James and went back to join the others waiting for the boss to return with the part they needed to get production started up again. I glanced at the set up as I turned to leave, and was surprised to see how simple it was. A few of the laptop cases were opened up with sub assemblies and a few prepared modules ready to put in place. I noticed there was one unit in each of the laptops that was already in its own separate case ready to be installed. That was the special module that had been prepared in Dallas. They would just drop things in place, screw it together and pop the case back together. Another laptop came off the line. James and Jeb had it all organized and had instructed the locals. The touchy nanotechnology had been done at the main branch in Dallas, and this was a very simple final assembly. I didn't believe it would be that simple and I was right.

It was another major shock for me to realize that these guys were operating completely within the law. I was the one who was out of line, charging in here the way I had done. I had to be sure my own attitude was right before I judged the behavior of others. But a lot of facts are needed to make proper decisions about anything. I walked back to my car, cranked the windows down and sat down in the driver's seat. I called Willie on the cell phone and he answered like he was disturbed about something.

"Ah, call yah back," he said.

In the glove compartment I found my box of cards, pulled out the one for Von Mueller and punched out his number. When he answered I told him about my confusion regarding the Gladstone super-laptop. I said that I had read some of the literature about nanotechnology, when I was at the Global Technology Conference. One of the handouts talked about the fear of runaway nanobots or gray goo that could eat up the biosphere. I told him it made me wonder what I was doing on this case.

"That just shows where we started, compared to where we are today, Dan," he said.

We now believe those gray goo type nanobots, which we originally feared, are not an essential part of the molecular manufacturing process and can easily be eliminated with the safeguards we now have in place."

"I read a recent book and it was about runaway nonobots that were self replicating," I said.

"Oh, well no wonder," he said "That's a great writer, and it was researched well. But the gray goo scenario is now obsolete. It's been relegated to science fiction."

"What about industrial accidents?"

"We know today, that gray goo can't be made by accident. If it could be made, it would have to be intentionally produced, and that's not going to happen."

Dr. Von answered all my questions and convinced me that my fears of the self replicating robots were not based on recent studies and should be put aside. I forgot about it and get back to work. If we could find Bob Clauson he could give us back our perfect example of the super-laptop that produced no gray goo. I still had those handouts about nanotechnology stuffed in the glove compartment. I left them there and started up the car.

Driving to the end of this block, I turned right at the traffic light and drove on around to the large industrial sized driveway in the back of the Comp-ease building. There was a six wheeler parked with the rear end backed up to the back dock. I parked and got out. I could see two smokers standing outside of the open overhead door. They didn't seem to pay any attention to me, so I walked over and took a look at what had been delivered. The trailer was full of boxes like the ones I had seen stacked on the floor by the assembly line tables. I thought that they had never really started production on the assembly line. For some reason James Lebeau had gone somewhere and whatever it was he went to get was needed to start production. The others had to wait until he returned, but then the process would begin and the trucks would be making round trips. This was a smart operation. I couldn't prove anything wrong, but I still thought that there was.

I returned to the car and my cell phone sounded off. It was Willie calling me back still excited about something.

"Where are you at, right now?" He asked.

"Strange as it may seem," I said. "I'm parked in back of Comp-ease in Luling"

"OK, Ray is parked across the street a couple blocks away with his glasses trained on you right now.

"That's perfect, Willie, so what's going on with you?"

"I followed the top boss at the factory the Lebeau guy to a house about a mile from there, and here is the thing, Dan. He ended up taking me to that same damned house where Oliver took us. The boss man parked in the drive in front and went to the back of the house and talked to a man who was waiting in the back of the yard where that big drop off is. I parked back on the curb down the street a ways and walked over to the yard next door. The man he talked to looked exactly like the picture you got from Gladstone of Bob Clauson."

"So did you get them?"

"No. The guy that looks like our Bob handed a few sheets of eight by eleven paper to the boss from Comp-ease and they both turned to go their own way. Well they both saw me when I came out of the bushes next door and went for Clauson so I guess you know what happened."

"He made it to the blue BMW before you could get him and you didn't want to shoot him."

"It wasn't a blue one this time but a red one. He ran real fast. I even tried to jump the drop off to catch up to him and busted my ass. I got one shot off to a rear tire but he was getting too far away then for me to risk another shot. Of course, by the time I got back to that empty house the boss from Comp-ease was gone too."

He had made a lot better progress than I did, given the same situation. I said for him not to worry, go back to Comp-ease with Ray, and that I would check with them by cell phone at least daily. I wondered about the red BMW, instead of the blue one, but I wasn't going to question Willie about it. He knew what he saw, and he would have mentioned it if there was more than one person in the car or anything else that I should know about.

I thought that the Gladstone crowd would never believe me if I told them Bob was the criminal behind all of this. Willie could be wrong. It might have been someone else, not Bob, as he had said at first. What was given to James Lebeau could have been electronic engineering drawings and written instructions regarding production of the new laptops. I was stymied on this one for a while and I was still wondering about the BMW cars of different colors. I tapped out the number for Wallace Norton and he answered with a relaxed tone of voice.

"You're not running to a classroom this time," I said.

"No," he laughed. "This must be Brandon. I recognized your voice."

"I'm glad you're doing better today and what I have to ask won't take long."

"No problem, plenty of time today."

"It's a bit personal. I've noticed a lot of people lately have been driving BMWs and I was wondering if there was a dealership around here offering good deals on them. Would you mind telling me where you got yours?"

"No, not at all, as a matter of fact I did get a good deal. I got it at Whittington Motors in Lockhart. Nelda's father owns the dealership."

"She probably wouldn't need to buy one then, he might just give her demonstrators to drive around in," I said. Blue and red were common colors in these modern times.

"Yes he does do that. I've seen her driving several different ones."

"She has been driving a red one I know. Has she driven a blue one also?"

"Yes, she does at times," he said. "I hope everything's going alright with Nelda."

"She's doing fine, no problems. I was just thinking I might get a deal on one myself. Say, I'm going to have to be the one to run this time, Wallace."

He told me to have a good day and I said I would. As soon as I closed with him, I grabbed Nelda's card and remembered she no longer had that number. And the license tag on that red BMW would be a dealership plate, registered to the dealership, not to the man who was using it. I thought about how Nelda had said Bob was like her brother, but that she had wished it had been more. She was in love with Bob, and whether this guy was Bob or someone else, she was involved with that person and she was letting him use her cars. But I had to check out my theory. I called Gladstone and asked for Marsha.

"Gladstone Electronics, this is Marsha Brooks speaking." Her soft voice came on.

"Hello beautiful, we haven't had lunch together for a long time."

"Dan, how are you? Nobody seems to know what you're up to.

"I'm doing fine. I may need to contact your new high-tech employee."

"You mean Nelda Whittington. Yes she's here today and I'll have her call you back."

"She's working in the lab, or you saw her?"

"Both, I saw her this morning when I came to work."

"You saw her in the parking lot?"

"Yes."

"What was she driving?"

"A blue BMW, Sherlock. You called for info, not for lunch."

"I called for both, but too late for lunch now, how about later?"

"OK, I think, but call me first. It's a real busy time today."

I said I would and closed my cell. Talking to Marsha about lunch reminded me that I had missed mine and was hungry. I wondered if James Lebeau had stopped to eat something on his way back to the factory, so I drove back to the front and parked next to the reserved space. That space must have belonged to Lebeau, because in a few minutes he pulled into it, parked and got out. I got out of my car and walked to the curb to confront him. He could tell something was brewing. I stood directly between him and the front door to the building.

"Come inside to the office and we can talk, Mr. Brandon." He said.

"You stopped for a bite to eat, on your way back and you're refreshed and ready to get back to work."

"Yes I did have lunch, but I bet you missed yours. Come inside, I've got some coffee and donuts to give you."

"No thanks," I said and continued to stand in his way. "I don't have the time to waste and you don't either. I have a short message for you and I think you should listen to me. Either on your person or in that car, you've got the instructions you needed to start production of the laptop. You got them from a man who met you at a house here in town or I should say outside of the house?"

"That was one of your detectives who attacked us and even shot at us."

"He didn't shoot at either one of you, he only shot at one of the rear tires in an attempt to stop the man in the red BMW."

"Why did he want to stop him?"

"He's a missing person, and is wanted for questioning by the FBI, and several Texas law enforcement agencies."

"I had seen him before around here. He claimed he owned a ranch in the area and was an electronics engineer who could solve our problem. He came to me. I have no idea how he knew about us and our problem, but he made so much sense that I decided to do business with him. I read the instructions and I know it will work. If it's evidence you can have it." He pulled out several sheets of paper that were folded up in his pocket and handed them to me. I took a quick look at it and didn't understand a thing about it. You would have had to be an engineer like James Lebeau.

"I'm going to turn this into the FBI," I said. "If you have any more evidence or information that will lead to the whereabouts of that man you better turn it over."

He said he didn't have anything else, and meekly asked if he could go now. I said that he would probably be visited by the local police and the FBI, but as long as he was cooperating with me I had nothing to hold him on. He went inside the building and I got back in my car. Before I drove back to Austin I had to take on some nourishment. I stopped at a barbeque restaurant and filled up my gut.

CHAPTER 16

The first thing I did, when I got back to Austin was to go to the Federal Building. I asked for agent Tom Cobb, and he met me in the same room as before. I told him about the latest developments and gave him the instructional information I had received from James Lebeau. Cobb already knew everything about the new Comp-ease division in Luling, except for the meeting between Lebeau and Bob, or the man who looked like Bob. He didn't seem to know if it was really Bob either, but he had the same wise, know it all look.

"You did the right thing to bring the evidence to me," he said. "I'll give you several copies to give the police in Crystalville and Austin. Tell then to check with me if they want to know anything more about the original and of course the same for you."

"I didn't have a chance to investigate things at the UT lab when Nelda had the Gladstone laptop delivered to her. They were so happy about it at Gladstone that they gave her a job."

He said he was aware of all of that too, and he wanted me to stay away from Luling. He left me with one gem about Luling, and that was that the name came from a Chinese worker on the first railroad. I let myself out the same as before, and went back to my portable office in the car. There was no time left on the meter, so I set sail for Crystalville at once.

The CPD building was located on Lake Crystal Drive between the old town and the new town. It was windy and several sailboats were on the lake as I turned off LH-1. I drove on Lake Crystal Drive to Online Avenue, where I turned left and parked at the rear of the building. I went inside through a rear entrance and easily found Chris Amani's office. I gave her two copies of the technical lingo and filled her in on all that was new.

"Is one of these for me to give to the APD?" She asked.

"Yes, unless you'd rather have me take it myself."

"Let me have one of your cards to give them, and I'll tell them they can check with you if they have any questions."

"I thought that would be best, since you've already been dealing with them."

"Yeah, you know how cops are," she smiled. "They want to be able to decide who they set up a relationship with."

I returned the smile and said, since most of them were males, they might prefer working with her. Chris wasn't at all vain, and she had a good laugh at that.

"One of those old boys has been a little over zealous," she said. "I've had to curb his enthusiasm." We both had another laugh and I was ready to change the subject.

"Has there been much progress in locating the rifle that was used for those murders?" I asked. "Since they know it was a lightweight Remington chambered for a 7mm bullet, couldn't their expert take any rifles we get from suspects into the ballistics lab? They might compare the barrel markings with the bullets taken from the victims."

"Yes, some of the truck drivers go on the road with a hunting rifle and we have stopped some of them and sent their rifles to the ballistics lab already. The APD has been doing it too, but so far we haven't had a match. If we get a match we've got our killer"

I told her that I had my eye on a few hunters and would be looking for an excuse to examine their rifles. The only trouble with that was the personal aspect of such things. You would need to have about as much evidence as you would need to arrest a person, before you could confiscate their rifle. She said for me to give her regards to Ray and Willie. I said I would and walked back to the rear parking lot and my car.

Driving back the way I had come I took Lake Crystal Drive for several blocks, but before I came to Lake Crystal I turned right onto Gladstone Drive and took it to the parking lot. I parked and went inside.

All I had to do was glance around at the lavish exterior and interior materials that were used in the construction of this building and the old paranoia started to return. But my profession was known as the paranoid profession, so I guess I was just about par.

Carl was standing outside of his office and he motioned for me to come over. He had a coffee pot ready and we both got a cup and sat down in the hard-backed chairs.

"Long time since you been around, Brandon," he said. "You know we got some kind of super-laptop to guard now?"

"Yes, it's supposed to be the same one."

"Yeah, I just got a replacement for Campos, and the other two had their hours cut drastically. Have you talked to them at CPD about those murders?"

"I just came from there," I said. "Chris said they've been checking hunting rifles taken from truck drivers who shouldn't have had them on the road."

"You think they'll find a match to the bullets from the victims that way, Brandon?"

"No, but I guess you do the best you can."

"The perpetrator's not some random nut, but a nut with a motive."

"Money and power are the motives that come to mind," I said.

"Sure, you know we're in a hell of a spot, it could be any of our bosses or their friends. They all own company stock and it goes sky high when their laptop starts to sell."

"Are you into the stock too, Carl?"

"Not enough to get rich on it if it goes up. We'd be giving ourselves a phony ego boost to include ourselves on the list of suspects."

"I'd have to agree with that, but just plain anger can be a good motive. My anger curve shot way up when those three worked me over. I had a good alibi though for the murder of Tom Oliver, since I was on the scene dodging bullets myself."

"Yeah, I noticed the APD didn't bother to call you in for questions."

"I know our suspects could also profit from being a turncoat," I said. "That super-laptop could have been sold twice already. For all we know Comp-ease paid a special figure to the special someone to just copy some things, and then that same someone turned around and bargained with Harold to set it up so Gladstone would buy it back again. So how would that be for getting a lot of mileage from the same laptop?"

"It would be good," Carl said. "But, from the standpoint of personal greed, if it's Harold or Von and they've been able to do all of that, maybe they might sell it again. Look at how the greed led the Enron CEOs to go so far stealing from their own company."

We were putting away the coffee like we were on a cheap drunk, but it was keying us up, not making us lethargic like booze would do. I thought the paranoia we shared was going to help solve the case. Before I got into the PI business and I had the time to spend studying the game, I had played some tournament level chess. I remembered how you got better by learning all the ways your own king could be attacked, and you protected your king while you made moves to endanger the opposing king. You didn't improve your rating by being well-adjusted, but you did it by being paranoid.

Carl said it was time for him to make his rounds. He said for me to finish my coffee, so when he got up and marched away I remained there for a few minutes thinking about my next moves on this case. We had been thinking a lot about motives, because it was a necessity on this case. We were still struggling to get some suspects and start collecting evidence against them. It did make it tough when you had to consider your boss a possible suspect. I had always subscribed to that theory of loyalty to your boss, and if you couldn't be loyal you should quit and go work somewhere else. I finished my coffee and walked back to the elevator and took it to the third floor. Marsha was sitting at her computer and she motioned for me to sit down next to her. She was working with a huge stack of print outs and I thought there was no way she could leave here to even take a break. You could see it in her face.

"Get some goodies, and coffee," she said.

"If you haven't had a decent meal lately, I'm your man," I said. "I'll take you away somewhere to never land and you'll be treated like royalty."

"Yeah, to 'Fish King,' or 'Hamburger King?"

"I had something more substantial in mind for you." I claimed with enthusiasm.

"Thanks for the offer, Dan. I'll have to take a rain check."

I took the corridor down to the big guy's office, and rapped on the molding next to the open door. He had his nose in a stack of paper and a pen in his right hand. He was about to write something and he looked up when I knocked. He motioned me to the chair next to his desk and I sat down.

"We're all busy trying to get our laptop into production, but we've always got time to hear how the search for Bob is going."

He sounded like an FBI agent. I was sure he was aware that the FBI was on this case now. I thought he must be in denial. He had deluded himself into thinking he could resolve the companies problems by hiring a PI to do it and still keep everything quiet. It had been a taboo subject with us, and he still didn't want to talk about it.

"We think Bob has been spotted and we're working on it," I said. "As you know, the FBI is on it now, too."

"Yes, and there isn't anything we can do about them being on it."

"All the police and federal agents have done well keeping it from the press." I said.

"Of course, I know you were right about us needing every resource we could tap into. We still need you too. Your agency has turned up everything the feds are working on now. You got abducted and beat up yourself in the process. Gladstone is grateful and so are the feds."

"That sort of thing is always nice to hear," I said. "And since you're in touch with the feds yourself, I'd appreciate it if you'd keep me informed of any news from them,"

"There is something, the document you rightly turned over to them has been sent to an FBI lab, to Quantico, I think he said. They compared the writing on it to the writing of a sample of Bob's handwriting we had here at Gladstone and it matched."

"Now we know for sure it's Bob. I have a man in Austin I've used before to check handwriting, but I thought it was better to let them do it all. When they get the results back about the notes and instructions from Bob, it will tell us everything there is to know about the paper, pen and ink used on the original. That's why it was so important for me to give it to the Feds. There are no other labs that can compete with their labs."

Harold Remkin walked me all the way out of the office talking to me like I was one of his oldest, most trustworthy friends. As we went past Marsha she didn't even look up. He stopped at the elevator and said he would let me know when he heard more from the feds. I couldn't ask him any tough, personnel questions at that time, but he was always just a cell call away.

As I got off the elevator on the first floor I saw Carl come out of his office. "I've got something to show you." He said. Following him out the front door and around to the rear of the building I noticed he was flushed and breathing hard. We turned and, walked to a spot on the ground. He cautioned me not to go any closer to it. There was a long hole connecting the wooded area behind the building with the rear wall of the building. Carl said for me not to slip my foot into it, as he had dropped a quarter into it and never heard it hit bottom. When I asked him if there was an explosion he laughed and said there had been a vehicle back-fire from the county road nearby.

"Don't you know it's getting like California around here with buildings trembling and the ground opening."

CHAPTER 17

Driving the shortcut back to my super-room at the Crystal Palace, I took Gladstone Drive to Lake Crystal Drive and the short loop to the front of the hotel. I parked in the space by my room on the first floor. In spite of all the suspicions I had about these people, and the paranoia that seemed to float along like fog on the lake, it felt good to get back to Crystalville. I could listen to the waves slapping against the shore and be asleep at once in this peaceful environment. Some paranoia would go with me anywhere, but that was my job.

By the time I got inside, and turned on the television low, I was ready to sit in the recliner and read awhile. I ordered a ham sandwich and coffee, through, my great room service. I said to just send one cup of coffee this time, as I was getting a bit too much lately and wanted to temper it down some.

Switching on the reading light, I leaned back in the recliner and browsed through the newspaper. Gladstone wasn't in it. I wondered if some of the big guys like Harold and Albert had arranged to pay people to keep the company business quiet. It happens often in our society. Money and power can be used to control public opinion. Propaganda can be disseminated to change our thinking, and pay-offs can be made to be sure we don't even hear about certain things.

Even with the volume control set low, I could hear the persuasive messages coming from the news channel. And I thought anyone living in these times would have to know that there had been a lot of political manipulation going on, regardless of who they chose to think was doing it. Lounging around in all the comfort afforded by my super-room, my eyes grew heavy and I started to doze off.

"A ham on rye here," Albert said as he rapped lightly on the door.

"Be right there," I shouted. Poor guy had no help to deliver orders. I didn't believe that for even a moment. He was an intelligent, well educated man, and I had been giving him credit for doing about anything, as one located in a seat of local power. Then why would he go to such trouble to snoop around? I opened the door and he brought in two orders.

"That ham looked so good, I had to have one too." He said.

I asked him to join me, since there was a small dining table and a couple chairs in the outer perimeter of the room. This was a special room for special guests like me to have some influential friends over for a drink or luncheon.

"Oh, I just put them together on the tray for convenience. I could take mine over to my place, I thought, and not ruin your privacy."

"Don't worry about that," I said. "I'm glad to visit for a while." I did want to take this occasion to investigate him further, maybe find out what kind of hunting rifle he used on the big hunt, with the other big boys.

He took his plate and a bottle of soda, put them on the table and sat down. I took my plate and the coffee to sit it down in my recliner where I was already set up.

"So how have you been doing?" He said. "You're always off investigating and Harold said you were here for a rest."

"Well, as you know there was that shooting accident I had to check into at the hotel."

"I wondered about that."

"Of course we're trying to keep that sort of thing from the public."

"Yes, Gladstone has to walk a thin line during all of this. As you may have guessed I have an interest in the company name also. Anything you tell me will go no further."

"I thought you'd be able to keep things to yourself and I need someone who does some hunting to give me advice."

"I can do that, sure."

"It was a deer rifle shooting I've been checking on."

"People get shot all the time in hunting accidents," Albert said. "And while cleaning their rifle."

Not telling him how my rifle accident had happened, I went into a lengthy conjecture about how I needed to know more about the various types of rifles, and how they were used for different types of hunting.

"There isn't much to it," Albert said. "It depends on what kind of hunting you're doing and where you are. If you're in a deer stand just waiting for them to come and eat corn a heavy thirty ought six is fine."

"But those things will kick you on your tail."

"That's right I wouldn't give one to my nine year old niece.

"What kind would be good for the mountains and other places where you might have to run them down?'

"I own three of them, and I have them stored in the lighthouse at the edge of the water. Do you want to see them?"

I said that I did and we took the short walk over to the place. He played around for a while and came up with the key to the door. He opened it and we stepped inside. There was a spiral staircase if you wanted to go up to the top. What we were looking at was like the inside of a storage shed. He said that he kept it locked and had several storage sheds to store maintenance tools in, so he used this one for some of his own things.

The lighthouse was only used to keep his sporting equipment. There were several cabinets placed around the periphery and the remaining wall space was taken up by special hunting items that were attached to the pegboard by means of hooks. There were two of the old standard thirty ought six rifles hanging on hooks, and there was a space with hooks for a third one that was empty. The silhouettes of the rifles were painted on the pegboard in white paint the same color as the exterior to make it stand out. At first he pulled back in surprise, as he looked at the empty space.

"Oh, I remember, I loaned that one to a lady friend," he laughed. "I guess this explains my point anyway. It's a light Remington that a lady can use without getting kicked on her tail. I'm sorry I can't use it as an example like I was going to do."

"That's OK," I said. "I'm familiar with that one, probably uses a three ought eight Winchester case, necked down to receive a seven millimeter bullet."

"Yes, exactly right and you seem to know more than I thought."

I was silent for a full minute before I responded, waiting to see if he would add anything to what he had already said. It was my intention to put a little pressure on him while we were on this subject.

"My general knowledge of weapons I gained from the military service is not too bad, and I have to know something about ballistics on my job."

"Then why all that talk about wanting to know more about deer rifles? I think you already know about them."

"I'm sorry about that, Albert, but I have to check out anybody who has deer rifles on this case and might have a possible motive to shoot people who seemed to be working against Gladstone."

"Shoot what people?"

"There were three men involved in the theft of the prototype-laptop and the abduction of Bob Clauson. They were all shot and killed with a deer rifle like your Remington."

"Wow, I get it now, so I'm a suspect?"

"You're not really a good candidate to do something like that, but I had to level with you about this. The Crystalville police and the State Troopers have been confiscating that kind of deer rifle to be checked out in the lab. They have the shell casings and bullets used in one of the murders."

"So you want me to get that rifle back."

"Yes, I'll drive you there to get it right now.

"But shouldn't I call first?"

"No, it's here in town isn't it?"

"Yes.

Albert went with me back to the front of the hotel where my car was parked I told him he had better get inside and give me directions to his lady friend's house while I drove. We took the road which looped around the south side of the hotel becoming Lake Crystal Drive. It divided the old town from the new town. The post Gladstone section of the town was south, which was to our left as we drive along. The avenues were given names related to computer technology. He had me turn left on Silicon Hills Avenue and take it to Drexler Boulevard where we stopped in front of a modest red brick house and got out.

It was turning dark outside and the lights had not been turned on in the front of the house, so we walked around to the back. There were no lights on in the back of the house and no car in the garage. Just as Albert announced that nobody was home, a new white Mercedes pulled in the driveway and the garage door rolled up. I thought the driver looked familiar as the car was driven inside and parked. When she switched on the outside lights and motioned for us to follow her inside I could see it was Marsha Brooks.

"I'd have never bothered you about it, but Mr. Brandon said there's something about that kind of rifle I gave you, that has caused the police to be calling them in."

"Someone broke in my garage and stole it last week. I was going to get you another one, Albert."

"Oh, nonsense you'll do no such thing," he said. "I hadn't heard about that happening."

"I didn't want to tell you, till I had a chance to get you another one. Well come on in the house and have a little drink with me, you guys, we can talk inside."

After making so much out of it all we had to go inside, be civilized. There was no way I could see Marsha as the shooter. I didn't even know if she had learned about the three related murders and I wasn't going to mention it. It would be easy to check about her break in and the theft. I was on good terms with the CPD now, as well as with Carl Watson.

She led us through the common door between the garage and the kitchen of the house. We went from kitchen to a combination dining area and entrance parlor. She made us comfortable in the parlor and brought us gin and tonics.

"How did you happen to borrow Albert's rifle, Marsha?"

"I've hunted with Albert and my Father since I was a child. When I reached adulthood, I rationalized deer hunting by remembering how the over abundant population of deer would die of starvation if they survived hunting season."

"Me and my big mouth is the reason she ended up in this position," Albert said. "We were talking about how she had not hunted for a long time, and I shoved the old Remington at her. It was the one she had used for many years and I thought if she had it she might use it again. I'm really sorry I did that Marsha, since I know it was my problem not yours."

"I don't think it was a problem for either of us," Marsha said. "Two years ago my Father died, Dan. He had been the one to raise me, since my Mother had passed on when I was seven years old. So, you can understand how Albert and I have missed him terribly. I think we've gone into role playing. Albert has always shared that father image with him, no wonder I think he's my second Father and he thinks of me as a daughter."

Here I was again in my own role, I thought. It was the role of the PI man who stood alone, and was never able to be very personal and human with others. I also thought that I should have been ashamed of myself, but of course that would never work for me.

We finished our gin and tonic and I had made two great new friends and possibly had lost them all in the same second, while having gin and tonics. The tears welled up in Albert's eyes and Marsha kissed him on the forehead as we retreated out the front door to my car.

CHAPTER 18

It had been a very sad occasion for both of us at Marsha's house, and I was not at all proud of my part in it. I drove Albert back to the Crystal Palace and let him out as soon as he got himself under control. I was wired now. I had been drowsy in my room and had left the coffee untouched, when we had tossed down our ham sandwiches and went running out into the night.

Driving the loop around to Lake Crystal Drive, I took it to Online Avenue and parked behind the building as I had before. I was shocked to see Chris Amani in her office. I expected someone to be there twenty four seven but not her.

"It's time for the night shift," I said. "You need some sleep, Chris."

"You're just projecting that to me, since you work too long yourself."

"Well, maybe so, but it takes one to know one." We both laughed. "Last week you had a B&E and theft of one of those perfect rifles for our serial killer."

"Oh, yeah, Charlie Sloan was called out on that one and he's here now."

I hung around till he came back in the office. He had the same relaxed manner and that half smile I remembered him for. We joked about how detectives often projected their own faults to other detectives and the case in point about me noticing how tired Chris seemed when I was bushed myself. However, I prevailed by persuading Chris to go on home to bed. She had already pulled out the typed theft report and handed it to Charlie for him to refresh his memory. It had only been a week so it was easy.

"Someone forced open the back window to the garage and climbed in from the back yard," Charlie said. "It was still open when I got there."

That was before the Gladstone case had even started," I said.

"At that time we didn't have three related murders with the same kind of rifle, so it was just another theft to me," he said. "I didn't know to connect it to our case."

"We only knew about one of them, and the lab reports weren't in yet," I added.

"The lady was talking with a lot of guilt, like it was her own fault that someone broke into her garage and took the rifle. She said she should have put it in an inside cupboard or in the kitchen where she would have been more aware of it."

"Marsha can never forget her responsibility to others," I said. "She has one of the top positions at Gladstone Electronics, and works with the executives every day."

"It wasn't like I put a star by her name for acting guilty. She reminded me of my wife." Charlie said. "Shelley always acts that way. You would think it was her job to absorb guilt, but it doesn't mean anything."

"That's what I think too, Charlie. It's a common personality trait to want to protect others from embarrassment. Marsha is a very thoughtful person."

"Yeah, I wish I could come up with something for us to follow up on but I can't."

The only thing I could think of was something I should have thought of sooner. How many others knew about the rifle in Marsha's garage? I could interrogate Marsha and Albert about it, but I thought I would put that off for a while. It didn't even seem very likely that the perpetrator had acquired the murder weapon that way. Maybe someone in the neighborhood saw her put it in the garage, and broke in and took it to sell for a little spending money. Neighborhood teen-agers were always in such need.

"Do you think I should interview her again now that we know more?" Charlie asked. He didn't just want to blow it off. I said it was a good idea and to let me know if he turned anything up. He said he would.

Just as I was thinking of leaving, Joe Allen came in and I got into another one of those three-way conversations. Joe was a deep thinker who shared my own feelings that motivations were very important. He also had his share of paranoia. We could be brothers.

"I'm still thinking about when we came on this case," Joe said. "You were telling us about Gladstone's commitment to nanotechnology, and how much money was involved in the special kind of research they had done to improve computers. Wouldn't some just loose it and start shooting, when they were presented with the prospect of loosing a good job or a lot of money?"

"Joe's been reading up on nanotechnology and all that molecular juggling around of atomic structure with big microscopes," Charlie said. "He thinks it's going to make us all come unglued and start a shoot out."

"What's the source for your reading, Joe?" I asked.

"It's the computer internet browser. Just type in nanotechnology and start from there. You get all the arguments put forth by Eric Drexler and all the other leading molecular research scientists. They all admit there's a potential danger coming up at some time in the future, as a result of the various nations of the world doing competitive research into nanotechnology. They're into the research now, so give them another five or ten years and the self-replicating atomic robots will be harnessed for warfare."

"He's right about it, Charlie," I said. "We all need to be interested and aware of what these modern day physicists and chemists are turning up about nanotechnology. Their jobs are as important as the computer engineers. They don't seem to think the danger is from industrial accidents, but from a purposeful type of molecular research that will allow some world power to use it like a nuclear war device, when it could be used to make the world better and a safer place to live in."

I added my own particular philosophy about it, which was that educating ourselves about nanotechnology was our civic duty. I could see examples of some very poor citizens every day on the television news channels. And here we were in a time when we should not only be good citizens of the United States, but also be citizens of the world. We shouldn't hide out from nanotechnology. Our country was already into it and we had to just be responsible about it. If we didn't face the issues that were here today, I thought, what would become of us in five more years. We could only hope that there were enough responsible people, in the other nations of the world, and that somehow we could all get together and prevent global destruction by a fascist nation. Nanotechnology that leads to molecular manufacturing would be used to the best advantage by democratic countries.

The three of us grew silent as Charlie got on the computer internet. He typed in google.com as the address on the browser. Then he typed nanotechnology for the Google search word, and we were impressed to see a full page of web sites related to nanotechnology come up on the screen. There were fifteen or twenty links to investigate.

"This is how you spread the word, I guess," Charlie said. He was clicking on one of the links as I left the CPD and went back to my car. It was the best example I could think of for spreading the word about nanotech. He had become a believer.

Driving all the way back to the end of Online Avenue, I pulled off on the last street before I came to Drexler Blvd. and drove till I could see the back of Marsha's garage and the back yard. I parked at the curb and walked through the grass to the rear of the garage. There was an outside street light on a pole to light up the area and I thought it was so bright that someone could have seen the crime committed. I could see where a knife blade had been used to pry apart the opening between the window sashes and to pry open the lever lock. It was locked now, but Marsha wouldn't feel too secure going to and from her car. She would be thinking someone had done the same thing again and was waiting for her in the garage, when she went to her car in the morning.

I punched in the numbers I had for the detective's office at the CPD on my cell. Charlie answered. When I told him of my concern about Marsha's damaged window latch, he said he'd call her in the morning before she went to work and say he'd like to come by with a maintenance man to seal that window up. I said I was surprised to see how well lighted it was behind the garage, and he told me that he had checked with neighbors, who had a view of that window from their own homes. None of them had seen or heard anything, but he remembered when he knocked at one house no one answered and he had thought he had seen some movement through their window. He would try that place again.

It was the quiet time when everyone has gone to bed that a lot of bad things happened. By the time I had been all around Crystalville several times and had made it a full night, there was still time for me to get a few hours of sleep. I drove back to Lake Crystal Drive and made the loop around to the front of Albert's hotel for special guests like myself. He was not lurking around this time and neither was anyone else, when I parked and went in my room.

A recurring dream invaded my sleep, a transportation nightmare where I couldn't get my car started. It had conked out early in the morning, when I happened to be by a car dealership which was just opening up. Everything else in the area was closed, and there were no people on the street. A pompous middle-aged man whom I, rightly, assumed was the owner appeared through a commercial overhead door which he had just opened. I couldn't imagine where all the customers had come from that early. There was a ring of them already following him around trying to ask questions about their own cars.

Joining the group, I started to follow him around with my own questions. However, I had to wait my turn and my chance seemed to never come. I woke up in a sweat and was happy to discover it was only a dream. Of course that's always the way with a nightmare.

I threw some water in my face from the tap, and heated up the coffee from last night in the microwave that was provided in the small kitchenette. Taking a quick sailor's bath and carrying my electric razor out the door I noticed it was seven o'clock. I traded the luxury of the super-room for my office in the car. Walking my fingers through the business cards I had saved in the check box mailer, I made a cell call to Nelda Whittington. After it had buzzed for about seven times a sleepy Nelda answered it.

"Good morning sunshine," I said.

"Oh, it's you, Dan," she shouted. "I recognized your voice."

"Have you heard any more from our FBI friend?"

"No, I only saw Tom Cobb that one time."

"He told me to stay away from Luling since the happenings there."

"What happenings?"

"They know all about it at Gladstone, so I thought you would know too."

"I know about it, but it's not what you think."

"I think someone has compromised Gladstone's super-laptop secrets for money."

"You could talk to Harold about that. The executives and board of stock holders have decided to share the nanotechnology with other companies. It's too big a thing for one company to sit on, and you could see all the problems it was causing. Gladstone is letting other companies buy into it now."

"What about the young man driving around in the BMW's from your family's car dealership?" The Brits with nanotech body and no petrol had sold more cars than ever.

"That's not what you think, either,"

"Tom Cob had evidence checked out in Quantico and it verified that it was Bob."

"I wish it was, but it still isn't right, Dan," she whimpered and then got under control. "Bob is still missing and that's why Harold is still employing you to find him."

"Yes and I'll continue to look," I said.

Nelda told me thanks and closed her cell. I closed mine too, but I was too stunned to do anything but just sit there for a while. That nanotech business was in a state of constant change and it was hard for ordinary people like me to keep up with it. I was going to have to remember that. There wasn't enough time to spend hours on the nanotechnology web sites every day, reading on the links about nanotech, but I could check with someone like Von. He had been glad to talk about the latest develop- ments and answer my questions. And for now it seemed like Nelda had got me up to date on it.

I had to walk a fine line between the facts and science fiction. People like me can do alright if they don't let their imagination get away with them, but I was thinking about the self-replicating nanobots I had read about. It got me started thinking about how the self replicating molecular particles had made a few copies of people in a science fiction book I had read, not much of a stretch from the actual cloning that is being done today. It was not hard to think that some sort of scientific phenomenon had produced a copy of Bob Clauson, and he, or it, was driving around in the Whittington Dealership's cars helping to start up new computer companies.

Breaking my focus from thinking too much nanotechnology, I started the car and drove around the loop to Lake Crystal Drive and back to Drexler Blvd. There was a white pick-up-truck parked in the driveway in front of the open driveway. Charlie and his friend had just finished the job on the garage window and were putting the tools back in the truck. Marsha had just come out of the house to thank them. I parked at the curb in front and got out. It must have been time for her to go to work, as she got in her car and started to back out of the driveway. We gave each other the friendly raccoon salute as she rolled on by me.

Although I hadn't shared in the work, I went up to lean against the truck with my comrades. Charlie introduced me to the CPD maintenance man, a toe-headed, light-skinned man, who laughed easily like Charlie. We engaged in a little chit-chat and I offered to drop Charlie off at the CPD parking lot where his car would be parked. There were a few other small PR jobs the maintenance man was prepared to do, so that made sense to him and he went on his way in the pick-up.

Walking out to my car, I told Charlie I owed him one. As we drove back to the station he insisted that those type of jobs were often being done by CPD, especially, when it involved employees of a powerful company like Gladstone Electronics. Charlie had smiled and mentioned something about Marsha being an attractive single woman for me. I assured him that nothing like that was going on, and that it was just a friendly business relationship. He didn't seem to believe it and I considered that to be a compliment. In about three minutes he was getting out at CPD and I had other places to go.

All of that business about Marsha had started me thinking about Claire. I had been missing her a lot, as I always did when one of us was out of town. There were always other women around, but she was the one for me. Ever since I had met her I believed in monogamy. She had made me the believer. Just on the chance that she might be back, I punched in her number on my cell phone and she answered by saying, "Hi Dan." She had caller ID.

"Good to hear your voice," I said. "I'm going to be working my way over to the Civic Center again, and I thought you might be around there."

"I'm at the apartment, just got in. Do you want to meet me at the Liberty for a short breakfast?"

I told her I would be there by eight o'clock and we closed our cells. Neither one of us was much for talking long on the phone. We needed to be available for our clients as much as possible because of our jobs, and we avoided idle chatter.

CHAPTER 19

It suited me fine to be getting out of Crystalville and back to Austin for a while. I had been in the doldrums and I was sure Claire would help me to get out of them. Driving down LH-1 for twenty minutes put me almost there. Another ten minutes in the town traffic and I was parked at the Liberty Café.

She was standing outside in the parking lot waiting for me, with that special smile. We embraced for a long time and then walked arm in arm into the coffee shop. With our vegan chocolate-chip cookies and a full pot of coffee we were ready for anything.

"I stopped at our office here in town and took home some things to work on prep for a case coming up," she said. "I'll use my computer to do most of the research."

"What I have to do is talk to Cal Denison, one of the exhibitors at the Global Technology Conference," I said. "I wanted to see if he knows much about the nanotech research that was done at Comp-ease."

"Is that another super-computer company?"

"They are now that they've had some help from Gladstone."

After I had related all of the latest facts to her about the case, Claire was as surprised as I had been, when I learned how this nanotechnology thing was larger than any of the individual companies.

"Maybe the president and his men are starting to regulate them some," Claire said.

"When you think about what all they put into the Iraq war, you'd think they'd be interested in a technology that could prevent the destruction of the world."

"Or cause it, if the wrong group is in control of the technology," I added.

"I read something the other day indicating that China and Japan were the largest investors into nanotechnology," she said with a dubious smile. Stopping herself, as she often does, she changed the subject and a more beguiling smile came over her. "We didn't know that we were finished with the other case till this morning, or I would have called you yesterday to tell you I was coming home."

"That's good, Claire, I thought you would have."

"Can you stay the night?"

"Yes, I'll stay it," I said.

"Yes, enough deep-tech stuff for now. Go see your man and come back to my place for lunch. I stopped and picked up some steaks. We'll have some fun, for sure."

Without even having to move my car, I walked to the Center's building next door, where I thought Cal would be tending his computer technology exhibit. There were several potential customers trying out some new business software, which might help them to make more money. Cal disengaged himself, came over to me and we shook.

"Good to see you again, Dan," he said. "I hope you haven't been under fire lately."

"No I haven't and how about yourself?"

"I'm happy to say it's just ordinary business that I'm able to handle."

"How did things turn out for you with the APD?"

"I leveled with them and it was just as you said. Of course, it helped when they found out that all that funny business with Thomas Oliver hadn't gone anywhere, and it actually resulted in our company having a legitimate business arrangement with Gladstone Electronics."

"I'm glad it worked out for you, Cal," I said. "Now, if you would, you could help me with something I've been struggling with."

"Certainly, after what we went through together, just name it."

"My boss at Gladstone has let me off the hook for the loss of the super-laptop, now that they have it back, but my job is still to find Bob Clauson. The FBI is on the case with me now. Everyone is looking for Bob and we still haven't got him."

"Yes, I've heard about him. My boss said it was Clauson, who gave Gladstone's laptop to him and James told the FBI agent. He said he paid Gladstone for the right to examine the laptop and then he gave it back to the CEO at Gladstone. You see, when he asked people at Gladstone about this man, Clauson, they agreed that he was the original inventor who came up missing, but they still didn't know what had happened to him."

The lawyers for both companies agreed it was a thing like no fault insurance. James Lebeau was in good faith and so was Gladstone. There were conflicting stories about Clauson. One of them was that he had help getting free from his captors and in reclaiming the super-laptop. He refused to return to Gladstone and some reports had it that he would be employed at Comp-ease, where he would lead the foremost electronic nanotechnology. Gladstone Electronics owned the laptop and it was returned. The question about it was how it was returned. None of the three statements given by those at Gladstone involved with its return seemed to agree. All Harold Remkin said was that someone called him on his cell phone and told him to send Dr. Von Mueller over to the parking lot at the University Research Center where he'd be given the super-laptop. Comp-ease was allowed to buy into Gladstone's new technology and then James Lebeau had to send the laptop back to Remkin at Gladstone Electronics. It was actually Clauson who was responsible for it getting back to Gladstone.

Nelda Whittington had said she got a phone call like Remkin's. Neither of them knew the voice of the caller. Nelda claimed a street person had delivered the super-laptop to her in the parking lot and then ran away from her. Von Mueller said Remkin told him to go over to the Research Center to get Nelda and bring her back to Gladstone. He laughed at Remkin not telling him anything about the super-laptop.

"That was about par anyway," he had said. "Harold never did tell me anything, till he had too."

When I thought about it, now, Cal Denison's statements were a little sketchy too.

"Did you ever meet Bob Clauson, Cal?" I asked.

"No, I never have seen the man."

"Did you deal with any of the nanotechnology aspects of your computers when you worked in Dallas at the main branch?"

"I'm an electrical engineer, but nanotechnology wasn't my specialty. To use that technology we had to come up with smaller computers. All of us were devoted to the improvement of computer technology. A super-computer used to be the size of an entire room and some of them still are, but the engineering research has been involved with bringing down the size of desktops to more the size of laptops. Of course, making a better laptop computer has been a good challenge for engineers, and those palm sized Blackberry's were just a result of some of the R&D we came up with on our way to designing computers small enough to work with a table top molecular manufacturing machine. We ended up with a real super-computer not much larger than a Blackberry."

Cal Denison didn't have any kind of explanation for the outlandish behavior of Bob Clauson nor should he have, since he had never met Bob. It seems that we foolish humans always want things to make some kind of sense. We always have a hard time accepting that if something happens it was supposed to happen. On the other hand, if you kept your emotions in check all the time and were completely logical you wouldn't be a human you'd be a robot. I wouldn't like that either. Modern people hired lawyers.

I couldn't help but think about the merits or demerits of robots, since they are into our culture in so many ways. Not liking robots is my problem. If I was an employer who had to worry about the work force, work conditions, and retirement a robot would be looking a lot better to me.

Walking from the Convention Center in a brisk March wind did something for my robot complex, so I continued on foot to Claire's apartment. She had a ground floor window open and I smelled the steaks at the entrance.

"The kitchen window was left open a smidgen and it must have worked," she said. "They're finished and I just now put them on the plates."

There was nothing to do but sit down and eat them, along with the delicious tossed salad she had prepared. Our steak eating rationalization was similar to the one that Marsha used, about how the deer would die of starvation if they were not finished off during hunting season. If we didn't eat steaks the cattle would be butchered anyway and others would still be eating them. And in addition to that, by eating steak we were helping to keep the Texas cattle raisers from going broke.

Some men have said they liked a good steak as well as sex. Well, I don't know if I would go so far as to say that, but Claire motioned to the books stacked on the computer desk and I gathered that she had a lot of research to do. I was so satisfied at the end of the meal that I stretched out on the couch and slept for an hour. I wakened to the sound of fingers on the computer keyboard. Feeling guilty about my short, refreshing nap, I told her I had to get back to work at the convention center, and would take her out for an enchilada dinner and a margarita when I returned.

It had been a startling discovery that a devoted research scientist like Bob Clauson could go through some sort of metamorphosis and come out of it as a ruthless, murdering money-grubber. You couldn't say he didn't have a good motive to shoot the same three kickers that had given me a sound beating. I was certain they had done the same thing to him. It occurred to me that since he had decided to join the vast ranks of the greedy, there couldn't be a better place for him to sell his nanotech secrets than at a Global Technology Conference.

Looking through the things I had stored in my portable car office, I found the original glossy photo of Clauson that I had been given at the outset of this case. There were fifty copies I had made at that time, and those that I hadn't given to my operatives were still in a leather-bound folder on the passenger side of the front seat. I grabbed the folder and went back to the building at the Center where I had visited with Cal Denison.

Cal was the first one I gave a picture of Clauson to, and he looked at it for several moments before placing it with other personal items on a shelf in his exhibit.

"In my opinion something put the guy over the edge," He said. "I mean it's such a departure for someone with all that education and ability to go wrong."

Thinking I had to make the most out of the circulation of this picture, I took the original out of my folder and held it up in front of us inviting him to take another look. I pointed out to him how the original was probably a bit more-clear than the copies. I studied it again myself, and mentioned some of the identifying features. He was an average looking man with dark-brown hair combed in a widows-peak and a solid straight nose. I thought that the wart located at the crease of the nose was more pronounced. It was on the left side of the nose, where you were supposed to put pressure to stop a nose bleed. The photo was from the waist up and in the left pocket of his shirt there was a case for holding some kind of glasses, probably reading glasses. Since his appearance was more like some movie star, like the young DeNiro, I thought if he wore glasses it would change his looks a great deal. I mentioned that fact to Cal and he responded.

"If he had glasses on he would look more like the physicist and electrical engineer that he is," Cal said. "You should point that out to others you might ask to identify him."

"Do you think he would try to disguise himself?" I asked.

"Yes, if he's the criminal we think he is, I think he might."

"You have to think like a criminal to catch one."

"That makes it even harder. As I said before his behavior has been so different than what you'd expect from a well established scientist that I can hardly believe it."

Contemplating how your own perception might be a hindrance to you, I tried to change my perception. I stopped thinking of Clauson as the special, electronics genius, and started thinking of him as a common criminal. I started drifting around the grounds of the Convention Center with a different outlook that would help me catch a criminal. As I scanned the area I tried to visualize Clauson in a disguise, more than glasses. He could have a beard, fake nose and more. I started thinking of how he might look in full disguise and I was more aware of the job I had to do.

I initiated a cell to cell conference with Carl Watkins and informed him about my new perception. He liked it.

"Yeah, a cop has to deal with different aspects of things, but you're right, Brandon, it's hard to do sometimes. In my mind I been seeing Clauson everyday, and I knew him a long time. I have to admit when I visualize him he looks just like he always did. Carl never thought about him using a disguise. The guy has such a common looking face I could never get used to the idea that he was a genius."

"It may have answered a question I've been asking myself a lot lately," I said. "Why hasn't someone spotted him yet? A lot of people knew him and he has had to have been moving around to have done all the things he's done." We ended our cell talk, but I continued to call others involved in this case and cautioned them about how Clauson might look in disguise. They all agreed there had to be some reason for him to be able to come and go so easily. The only ones to see him, since he had disappeared from the scene had been James Lebeau and Willie Harland. Clauson had been doing business with Lebeau at that time, and I had to assume he was driving around in one of Nelda's BMWs, because his classic Chevy was still parked in front of his Condo.

I made a cell call to Chris Amani at the CPD to see what she thought regarding the likelihood of Clauson having access to the BMWs. I thought Nelda would have known he was driving one of her cars and wanted get a second opinion on it.

"There could be a different explanation about the cars," Chris said. "He could have known she had a dealership car and tracked it with a remote global positioning system. She might not even know he was doing it, if he put it on the car at night. She goes driving out first thing in the morning, and he picks up the signal from the transmitter unit he attaches under the car the night before. The remote GPS satellite antenna is visible from the rear of the car. That's why I say he would have attached it after she had come home from work and gone to bed for the night. That way when she went out in the morning he could be anywhere in the general vicinity and just track it to wherever it goes without having to follow the car. He could wait till she got where she was going and then take the car. She comes out to get her car and it's gone. Those remote GPS units are available to the public, but we don't use them much, because of ethics. We have to have a very good reason to use one."

"It's an ethical thing with me," I said. "I'd enlist the help of the police, before I'd attach a GPS to someone's car."

"He may have only done it once or twice, and she didn't want to report it as a stolen car because she thought it was him. I'm just saying there are a lot of possible scenarios."

"The first time it happened it was a blue BMW, but we didn't know who was driving it. The second time we had two witnesses who saw Clauson in the red one."

"We could have an all points out on red BMW's with dealer's plates, Dan."

That was enough of an improvement for me to stick my neck out and I said yes. It probably was still the red one. It would beat what he had been driving before he thought of Nelda's family dealership, and that would likely have been a stolen car. He would feel safer driving his girl friends car, especially after he saw she wasn't going to turn him in.

I went back to meandering around the Center with the well disguised image of the genius engineer in my mind. Extending my pedestrian patrol to include the small shopping area and cocktail lounge, in the Hotel Complex, I was not surprised to discover there were no red BMWs in either parking lot. By talking more to the participants of the Global Technology Conference, I turned up more people who had known Bob Clauson. I started to hand out more of the pictures and collect anecdotal stories about him. Small groups of two or three were particularly rewarding to my information gathering technique. One man said Clauson had talked to him about new concepts of nanotech regarding how molecular materials reacted very differently in the nanostate and many things became possible through nanotech that were not possible before.

"He carried it a lot further with me," an older man chimed in. "He told me how some nanoscale materials could be given thin coats of experimental materials until the human eye couldn't see it. There wasn't anything he couldn't do with nanotech."

CHAPTER 20

Clauson was held in high esteem by his colleagues. I had the feeling that I could be wasting time. They weren't going to help me find him, if I was going to put him in jail. He was a folk hero. I thought they would take him in their own homes to give him refuge, as they spoke of him with awesome tones in their voices. It even had me accepting the possibility that through his special nanotech abilities he could make himself invisible. One lady physicist told about working on a lab project with him. They studied how various materials developed new properties when changed to atomic or molecular states. She said that he was so much into it that everyone around him caught his enthusiasm about making new discoveries. The way she put it was that if he told her a spaceship had landed, and the aliens were accepting him and one friend to go on a trip to another planet to study nanotechnology, she'd be begging to be the one to go with him.

"I'm the type that won't get on a plane," she said. "I drive my car everywhere, and we have nothing going on between us. But when you're into nanotech research with him it's a spiritual thing. He knows what he is doing every minute and you just believe in him. His light-weight nanotech materials allow more to drive big cars than did before."

That kind of talk had me stymied, so I stopped talking and abandoned my pedestrian patrol of the Convention Center. I was walking back to the coffee shop, when my cell sounded off. It was Willie and he was excited about a caper he had just completed that involved Tom Cobb, our friendly FBI agent.

"Cobb has had the air above us full of helicopters," Willie said. "They've been scouring the area for a couple days now, and about an hour ago there was a sighting of Clauson. Cobb took off in his FBI radio-car. I followed him to an old safe-house ten miles northeast of Luling. When you get to Luling you turn right on Pierce Street headed west that's Highway 90. It's also called Gandeslough Road. Get off of 90 on the west side of the San Marcos River on a gravel road. It will take you to a small two and a half story ranch house. That's where I'm set up watching the house with Ray and Cobb's driver, who's, already, filled his pants. Cobb told him to stay with the car and walked on inside the house. Cobb's still in there and we're just standing by. It's quiet, like before a tornado hits. Are you coming?"

I said that I was and ran the rest of the way to my car, by the coffee shop. I made good time on Hwy 183 and came into Luling on Magnolia Avenue. I turned west on Hwy 90 and turned off onto the gravel road in just a few minutes. I was driving about due north on it, when I spied the white, Victorian style house on the left bank of the river. It was in a state of disrepair. He said some of the siding was rotting and falling off.

Pulling off into the grass next to Willies car I took two extra clips for my S&W Pistol and put them in my pocket. It was quiet, alright, but nobody was there by the cars. I glanced down on the seat next to me at the MAPSCO page for South Central Texas. I noticed how the area between Luling, Segin and San Marcos formed an equilateral triangle. I laid my flat, metal, six-inch rule on the page and drew in lines between each of those cities. I had seen these geographic phenomena before, I called them trouble triangles. It was easy for people like me to believe such things. After all, I even believed in the invisible man.

It was windy and there were no rafters on the water to my right. The river roared on by like a beast of prey. I walked at a safe distance from it heading northeast on a path that seemed to wind around the small hill where the house had been built. It was spiraling up at an alarming rate when I came to a gulley that produced a flat spot to stop and rest a few moments. Something moved in the wooded area at the far end of the gulley, and then I heard a man's low voice trying to tell me something. He tossed me an object and I picked it up as I moved closer. "Do you have a gun?" I asked.

"No, I'm just a driver," he said. "Your friends are above us and they said to give you the bull horn in case you decided to use it."

I thanked him, and used the clip on it to attach it to my belt. I pulled the pistol from its holster, racked the slide and charged up the hill. I hadn't gone far, when I could see them ahead of me with Willie on my right and Ray on the left. They were both still about thirty feet below the foundation of the house and about that same distance below the skyline, waiting for me. It would soon be dark and I wanted this over before then. I motioned them in position farther away from me as we climbed. When we were only a few feet below the foundation and almost at the skyline I unclipped the bull horn and held the talking end up to my mouth with the left hand and held the pistol in my right.

"You have a federal officer with you in there," I said "You should release him right now and come out with your hands raised. The house is surrounded come out now. There is no way you can escape."

After pausing for a full minute there was no sound from the house and I stood up to start the charge. A shot from above sent a bullet into the ground in front of me and I dropped flat, tossing the bull horn to give myself a free hand. Willy had moved farther around the hill to my right and was out of sight. He had gone to a spot where he could see the rest of the way around the hill to where Ray was stationed. There was no way the deer hunter was going to get around us but he was slowing us down. The first shot must have been a warning, because he could have hit me easily at that range.

I could hear Willie running a short way and then another shot. Our rifleman was moving around from one window to another and he must have spotted all of us.

"Well, I'm hit in the arm," Willie said. "I'm wrapping it up and I'm alright to watch this back end of the house if you two can go on in."

"If you're ready Ray, we go on three," I said. We had all agreed on the code to go on a quick two count for such a situation. Actually, I went on one and Ray followed me. I went all the way to the front door before I heard the next shot and Ray dropped to the ground and rolled back down the hill enough to be out of the line of fire. I thought I saw the window where the round came from and fired three shots into it while I backed down to tend to Ray. The bullet had entered at the top of the left shoulder and did not exit. It would have to be dug out later. Ray claimed it had not hit bone. I took off the clean T shirt I wore under my sport shirt and pushed it over the wound and Ray put pressure on it.

I thought the shooter must have moved from the second floor to the first to be able to catch me coming in. I ran for it saving the remaining rounds in the clip. I was surprised that there were no more shots as I went through the front door. I went into the typical stance with the two handed grip, pistol out in front of me, checking all the rooms, working my way to the rear of the house. I heard someone making a grunting noise and went to it. Cobb was hog tied with masking tape over his mouth. I pulled off the tape quickly and continued to look around.

"Look in the cupboard," Cobb said. "He must have gone into a hidden passage and through an escape tunnel."

I took a kitchen knife, cut him loose and removed the blindfold. "He's going to be gone now, but you can have the honors of following out the tunnel," I said. "I have two wounded men to get to a doctor." He looked sheepish, grabbed his gun from the floor and ran into the passage.

It was a common thing for the early settlers around the Luling area to have those secret passages with a tunnel leading to safety. They came in handy, when the family was being attacked by an enemy. Luling had the reputation of having the worst gun battles in the nation, since it was on the Chisholm Trail, which was a main route for the cattle drivers. Wasting no time I hurried to help Ray and Willie into my car. I had seen a doctor's office on Magnolia Avenue and the bleeding seemed to be under control for both of them. But, still, I couldn't believe the misfortune of having two out of three wounded.

Driving faster than I should have, we made it back to the city and were heralded into the office of an elder-doctor by the name of Robert Hawkins. We went in waving our IDs at him. The doctor was gray and bent over some, but fast and efficient in spite of it. He pulled the bullet out of Ray easily, cleaned up and put a bandage on him.

"You can go back to work if you're careful and have it checked in a few days," He said. Turning his attention to Willie, now, he found the bullet lodged below the deltoid muscle, and definitely into the bone. "We have a good man here at the hospital who's a bone specialist."

"I don't need any hospitals," Willie said.

"Let's put it this way," Hawkins said. "If you don't want to loose the arm, you need a bone specialist. "He called the City Hospital, found the man was there and said we were on the way. It was only a few blocks to the emergency entrance at the rear of the hospital. We were met by attendants, who immediately put Willie on a bed on wheels.

"I didn't have much of a choice, did I." Willie said, as the attendants wheeled him away to the ER, where the specialist awaited him. Ray and I were asked to have a seat in a small waiting room. The doctor would talk to us when he was finished and Willie was assigned to a room.

It turned out that we waited about two hours. There was splintering and a badly fractured arm that would mend in time. They gave him an anesthetic and did surgery. He was still groggy when we saw him in the room. The doctor told us he would be there at least a few days, more if the re-examination showed he needed it. I thought Ray should stay with him at least until he woke up. Ray had an adequate amount of expense money and we decided that after he talked to Willie he would call a cab to take him back to the ranch house to get his car back and drive it to his hotel room. He would go back to the same kind of surveillance he had been doing in Luling and keep up with how Willie was doing too. Now that Cobb was aware of the men I had left in Luling, I didn't think he would object. Not since we had rescued him and were the only ones who could say he had botched up the job.

Cobb hadn't been the first FBI agent to go into a crime scene alone, so he could make the arrest himself and get all the glory. In his behalf, I could say that there was always a lot of competition between agents, and he may not have trusted anyone else to have gone in with him. The more people involved in an operation, the more chance something could go wrong. If you knew your own limitations and phobias you might be a bit ahead of things, but you never would know how others were going to react around you and that could be a great concern to an FBI agent with a control freak personality. I ended that line of thought, thinking that if you were an agent you had to be in control.

I told Ray I would call him later to see how Willie was doing when he regained consciousness. I went back to my car and was ready to start it and drive back to Austin, when I remembered that I told Claire I would be back to take her out for our evening meal, and I called her. After I related all that had happened to us on this case recently, she wasn't still mad at me, but was upset about it all.

"You said you would be back to spend the night, Dan, so I was worried about you."

"As soon as Willie had called me about the situation in Luling, I focused on that and forgot everything else." I told her I was really sorry about it, and she turned back into her old self. She gave me a wise crack about how cops could never be trusted, to which I replied neither could lawyers and everything was fixed. Before this time we had always been careful not to plan anything ahead. It had always been, "I'll see you when I see you." Claire said we had better go back to that and I agreed.

"After what you've been through, you should get a room there for the night and check back with your friends at the hospital. You were rushing things about your job because of me. Remember, we said we'd never do that, Dan."

"Yes, you're right, Claire, and I've been a dumb jerk as usual."

"Yes, but you're my dumb jerk."

"Thanks," I said. "I'll see you when I see you."

She said the same thing to me, and we closed our cells. I thought thank God for cell phones and mine rang again. It was Tom Cobb and I wasn't quite as thankful for them as I had been before. His kind of FBI efficiency had put Willie in the hospital and curtailed Ray's activities for a while too. He was sputtering something about owing me one. I told him about Willie and said for him to visit Willie, if he thought he owed us something. I gave him Willie's room number and he said he was going over there now.

"I'm in a visitor parking space in the back by the emergency entrance," I said. "I'll wait for you and we can go in together."

"Oh, that's good, thanks," Cobb said.

In a few minutes he drove up in the FBI car, a late-model, black Lincoln. He was alone, and I could understand why he would give the driver some free time after that ordeal. The frail man had signed on as a driver and had been given an active role at the crime scene, to pass on the bull horn and some information to me.

Cobb had me come over to the Lincoln, and I got in the front and sat down.

"I wanted to tell you where the tunnel went," he said. "It was a deep one and came out on the bank of the river, with a small dock for a boat. There was no boat for me to follow him in. Using the singular is wrong, as there were three shoe prints on the tunnel floor. I was trying to take him alive but it didn't work, so I went back through the tunnel and called in a helicopter with three trained men to comb the entire ranch inside and out. They are doing that job now as we speak."

"You went in alone, because if you called the copter down you were afraid it would cut down your chances of taking him alive, but it resulted in my man coming close to loosing an arm."

"Yes, and I'm very sorry about that. If I had it to do over again, I'd call it down and go in with three men as the original plan called for."

"I'd like to know your take on Clauson, since you've met at close range with him."

"He didn't seem very rational to me, not a highly educated man of science. That was the mistake I made. I said I would go in with my hands up if he would consider my proposal, which was to guarantee him his day in court and to assure him that he wouldn't get shot in the back. They said for me to come on up to talk, but as soon as I did they threw a bag over my head, tied me up, and gagged me. He didn't allow me to talk and he didn't say anything either."

I told Cobb about what I had learned after talking to the man's colleges, and he thought we couldn't take any stock in what they said because they were so biased in his favor. I didn't mention the nanotech possibilities for invisibility that had been espoused.

"Here is something you really need to know," Cobb said. "No matter what you and I think about this entire situation involving Bob Clauson, there are a lot of people in high places who want him protected at all costs. They say that what he can do with nanotechnology is so important to the world, that if he has gone insane and is a murderer we are going to have to overlook it, and be sure to take him in alive. It's a question of being needed so much that you're above the law. The leaders of every country in the world know there are always some that fall into that category of above the law."

"It sounds like your superiors in the FBI have briefed you to look at it that way."

"Look, I'm confiding in you and returning the favor. I don't know what he was going to do with me, he may be a psychopath."

The logical assumption seemed to be that a murdering psychopath could also be a renowned scholar and nanotechnologist, essential to the well being of our nation. I just had a problem with including that logical assumption into my own list of them. Cobb was telling me I better start thinking that way for my own good. I suggested we go inside the hospital and see how Willie was doing.

Hospital rooms were supposed to be quiet, but we were met with uproarious laughter as we entered. Ray and Willie were taking turns telling jokes to an attractive young nurse. Sometimes the sting you feel after you've been in combat causes a compensatory reaction. When the troops go home from war, there are a lot of babies born. Ray and Willie had wounds to remind them of what they had just gone through and were engaged in a bit of harmless hysteria. The nurse was aware of that sort of patient reaction and she was allowing it as a type of therapy.

"Now you boys settle down a little, your friends are here to see you," she said as she hurried out of the room to help other patients.

Cobb glanced at me, smiling in acceptance and I did the same.

"Did you hear the one, about what the homicide cop said to the lawyer?" He asked.

Willie was wiping his eyes, and said. "Yeah, we heard it, and it's time to settle down some like the lady said. It's good to see you guys."

"We got on a story telling jag there for a while," Ray chimed in.

"This is agent Tom Cobb," I said. "He's the FBI agent assigned to the Luling area."

"Yes, the very same one whose ass you just saved," Cobb added. "I don't know if I'm worth it, but thanks anyway." He went over to them both and they shook, while trying to curb their own laughter. Cobb seemed to have some of his own hysteria and the three of them had another short laughing jag.

It was obvious that we were all going to get along on this case. I had waited to learn all I could about Willie's condition before I called Helen in San Antonio. While he was talking to the others, I made the cell phone call to her. She was taken by surprise, as any woman would be, when she's been told that her husband had been wounded, and was in the hospital. I handed the phone to Willie and he assured her that he was fine and that she didn't need to make the trip because he would be home in a few days anyway. I got the name and address of the hotel where they were staying. It was the Heart of the West Motel on Highway 80 about a mile south of town.

Since I had decided to stay overnight to find out who owned the ranch property, I said I'd take Ray back there now to get his car. Cobb gave me a printed card with code numbers on it. He said the place was hopping with several copters full of men searching under large intensity lights. The card would identify us as cooperating local law enforcement officers. We could come and go as we pleased and ask pertinent questions. He said we could still use it tomorrow as well.

When I told him I planned to look up who the owners were he said he could save me the trouble.

"It belongs to a couple who lives in San Antonio," he said. "It's family property owned by Thad and Wilma Clauson."

"Is it Bob's parents?"

"No, it's his grandparents. His parents own one, north of that older one. It's on the other side of the black top road in the less primitive area. That newer safe-house is actually owned by Bob Clauson. Some of our men in a helicopter said there were two men who went into this older one with Clauson before I went in. It's hard to hear shots under a helicopter engine. They thought Clauson and his friends ran off three other men that had been inside the house. As they came running out of the house, one of the men with Clauson went after the unarmed men with a rifle. They must have escaped because the man with the rifle came back inside in a few minutes. If the helicopter observation was correct that was about the time I called to them and walked into the house. As I said, I didn't see anyone inside, but I saw Clauson and two other men go in the safe house before I went in."

Ray needed to get back to the motel and get a good night's sleep. I could use it myself for that matter, so I planned to return the next day to examine the FBI findings.

CHAPTER 21

There was a restaurant next to the motel and after I had shaved and freshened up I went there for some bacon and eggs with toast and coffee. The time of seven thirty is considered to be a late call for most PIs, but it was two A.M. by the time we got to the hotel and I got checked in with my own room. I told the desk clerk about what Ray went through, receiving a flesh wound and his orders from the doctor to take it easy for a while. I said I was his boss, asked for them to be sure not to wake him, but just let him sleep as long as he could. I also had the clerk make a note on his card that he had a free breakfast at the restaurant next door.

Not expecting to see Ray for a while, I drove back to the Clauson, ranch property. I parked on the grass at the same spot where I parked the day before. Men wearing FBI vests were picking up paper and anything that looked like a personal item and putting it in plastic bags. Most of them had a long stick with a nail in the end of it to save their backs. I waved my coded card at one of the men. He glanced at it and smiled.

"You want to go up to the house where the central point is set up," he said. "They go through everything up there, record it and keep anything that could be evidence."

"Have you seen agent Cobb here this morning?" I asked.

"No, but agent Stark is up there and he can answer any questions you have."

Standing there for a few moments allowed me to glance around and get my bearings, instead of just charging up the hill the way I had done before. I could see another small ranch house about a hundred yards farther along on the gravel road, and I started walking that way. The road curved and as I walked along I could see there was a clearing on the right side of the gravel road on the Clauson property, and there were log steps set in place going up to the house. We hadn't noticed the steps, because we had a siege mentality. We had pulled over in the high grass, parked and hurried up the hill as soon as we had seen the ranch house located west of the San Marcos River. I thought about fools rushing in. That's why they call it a siege mentality, as a warning to all of us fools. If you looked around a while, before you made your move, you might find a better way.

At the top of the hill I rapped on the primitive front door, as there was no door bell. A voice from inside said for me to come in and so I flung it open like I owned the place and went inside holding the coded card in front of me. "Are you Agent Stark?" I asked an older, bald-headed man with a handlebar moustache. He sat behind a card table in the center of the room.

Yes Sir, I'm Stark," he smiled, reading the code. "It says here you're a number one in the interested party category. I'd suggest you sit down in one of those straight-backed chairs till we get the last batch of the stuff and go through it. It won't be long now."

In only a few minutes several more bags were brought in and the agent who brought each bag went through the contents of that bag and recorded each item. Agent Stark collected the list of items and the contents of the last few bags. They were placed back with the others piled up in the corner of the room. Each of the bags had a number on it. Stark was scanning his list of the items that were just brought in, and placed a couple of marks on the page.

"And you're a private investigator by the name of Dan Brandon." By the way, Tom gave you a good rating too. We don't do everything in code yeah know."

"I'm glad to know that,"

"Well, so the house and the passage way tunnel were the only places we got anything. We got the shell casings left in the house and shoe print impressions left in the damp floor of the tunnel. Tom left one too, but his size is a ten and the other two sets were size elevens. We made a copy of all of the impressions and there were three sets of them. One set for Tom Cobb. He had been surprised to see the three sets of shoe prints there, before he went through himself. Since he was blindfolded from the rear as soon as he went into the house, and only one person had spoken, he had wrongly assumed only one person was there. There was no boat left for him when he came out of the tunnel and the shoe prints ended. We thought the three persons must have found a boat there and used it." "Cob thought that the person who tied him up was Bob Clauson," I said.

"It was so narrow that Clauson and his two CIA guards walked single file in the tunnel. We could see clearly, so don't worry—he was roughed up but he'll be O K.

Listening respectfully as I always did, when an authority was talking, I waited to be sure he was finished with his tirade. We can all get carried away with a thought. I'm no exception, and I think that helps me to be patient with others. What he said calmed me quite a great deal.

"You'll be interested in this, I know," he continued. "Tom told me about the case the two of you have in common, and how the shell casings left in this house matched the ones you and the CPD found at the Global Technology Convention murder scene. They were a 308 case necked down to fit a 7mm bullet. When they compare the markings on these to those from the other murder scene they will probably say there's a match and that all of the shootings were done with the same rifle."

"Yes, the ballistics report from Quantico will probably verify that," I said. "I wondered what Cobb would think about where our man went to from the boat dock. If his car was parked on the other side of the river, he might have just turned the boat loose from the other side and let it float on along with the current."

"We had it checked all the way to the end by helicopter, but no abandoned boat was spotted. There was no evidence that a car was waiting on the other side of the river either."

"If he didn't get away by car, that would mean he either went off on foot or took the boat to another place along the river and beached it, to make it harder to find."

"Believe me. Tom called the troops out to check all possibilities."

It was a certainty that he had men circling above the river and on the ground. Agent Stark gave me his card with his cell phone number, and said for me to check back with them. That was a lot better than a promise that they would call me. When people said they'd call you they almost never did.

I thanked Stark for telling me about the findings, let myself out, and went down the log stairway to my car. Climbing inside and starting up, I backed it out to the gravel road, but instead of turning right to go back to Highway 90 I turned left to follow out the gravel top. I wanted to check out the other small ranch houses and see where the road came out at the other end. In three miles it turned into a blacktop road with more elaborate ranch houses. I didn't think Clauson had gone this far on foot or in a boat. I turned around and drove more slowly looking carefully at the houses and the grounds. There were signs of occupancy at all of the other houses in this area. There was nothing out of the ordinary to make me suspicious, and I started thinking I was wasting my time again. I might as well be talking to the man's friends about the likelihood that he had discovered a liquid to spray on and become invisible. They had all said it would have been easy for him to do it. I likened Clauson to a foreigner who came here committed one or more crimes knowing he could claim immunity. Cobb had intimated that Clauson might be above the law.

Clauson was skilled in the disciplines of physics, chemistry, and electrical engineering. It was the perfect combination for a research nanotechnologist. His primary task was to come up with some more efficient new chip to replace the CMOS technology and he had done that. He had already designed microscopic chips with a new material that was dense enough to support better control of electron flow, with less current leakage and heat dissipation. This brought about the achievement of much less power consumption. A small amount of power could do a much bigger job with very little wasted energy through current leakage and heat dissipation. I had always heard that when electrical components got heated up there was a loss in efficiency, so it wasn't hard to understand what had been done. Electronic devices had been improved to a maximum in efficiency. I was aware that there had been teams of these nanoscience researchers working on it in universities all over the country. I wondered why we always had to have heroes to rally around. Bob Clauson had been portrayed as the fearless leader of nanotech and we all had better stay out of his way.

To face reality wasn't an easy thing to do, when you were looking at the kind of reality that I was becoming aware of. If I ran across this leader of the new world of nano-robotic-technology I was supposed to put down my gun and try to reason with him. The only other choice I had was to avoid any close contact with the man. That would be like bird watching. You could keep him under constant surveillance with high powered field glasses, but if he knew you were watching he'd fly away, become invisible or whatever it was he did.

This was the kind of game you were supposed to loose, so I guessed that I was doing a great job. If I could ever find this man again, I would have to give him to Willie to keep under surveillance. I wasn't good enough at that kind of long-range work. It was Willie's specialty and I would have to keep him available for it. So, even though it seemed like I was a looser, I had to remember what all was at stake here, change my strategy from time to time but never give up. For those of us who had become aware of it, we were into a new age, regarding the world-wide geopolitical situation. I had a job I could not quit, because it would be like quitting the world.

As I drove past the Clauson property I noticed the last few cars were leaving and the FBI search had officially ended. It was a wide enough road that I was able to just pull over to the right and sit there without obstructing traffic. I was thinking out new strategies, when Ray drove by and then backed up to park in front of me. He got out and walked back to me. I had the windows open and the doors unlocked.

"I hope you had that good breakfast like I did," I said to the window on the passenger side as he opened the door and got inside.

"Sure did, but they didn't give me the wake up call."

"You were supposed to take it easy for a while, remember."

"Yeah, I remember, but I thought I had been doing that."

I told him sleeping in, and having a good breakfast was a part of taking it easy. He wanted to know what they found in the search, and I related everything I could remember about my visit with agent Stark. Ray did seem rested, but I still wanted him to just sit in his car on the job for a while. That was part of my new strategy.

"Recent experience and some good advice from the FBI have been changing my attitude about this case," I said. "We aren't supposed to have any more shoot outs with Bob Clauson."

"Why is that?" He asked urgently. His light-brown hair had been growing out and he raked the fingers of both hands through it in an agonizing gesture. It was a bit disturbing to be told not to do that any more, when you had already earned your red badge of courage.

"It was my call and my mistake," I said. "Agent Cobb said he had put his gun aside to plead with Clauson to give himself up. Evidentially, Clauson is a necessary part of our national defense."

I went more into detail about Clauson's disappearing act from the boat dock on the San Marcos River, and presented my theory that he might still be near.

CHAPTER 22

Since it seemed more likely that Clauson would have found a place to hide in the older ranch house section, south of the black-top, I thought it would be good for Ray to go over it again. I assigned that area to him and drove my car further north on the black-top. My plan was to rejoin him in that section after I had at least made a cursory search on the road heading north and west of the river. I drove about three miles north till I came to another black-top crossroad. Turning left on it I drove deeper into the center of the trouble triangle and I could see that it was a vast area west of the San Marcus River. I I followed out that road till it came to another crossroad and by turning left on it I was headed back south again. When I came to the gravel road I decided to go back on the improved road headed north and learn the rest of the layout while I was there. It occurred to me that Clauson might have come out of the tunnel at the boat dock, untied the boat to make it look like he had made his get away on the river. If he had gone north on foot, and someone had been waiting for him in a car on the improved road, he could be in another county or even out of the state by now. I drove back to the first crossroad turned right and it dead ended at the river. Turning around at the end I drove west all the way to the other end of it, where it dumped into Highway 123. I had the feeling that he knew this territory and was still around here. By the time I had driven all the improved roads l was familiar with the territory too. I even discovered a well stocked store by the name of Clauson's Country Store and it had a small restaurant in it.

After driving around in the newer homes section for several hours I was ready for lunch and I thought Ray would be also. I located him with the cell phone and gave him directions to the country store. He was on his way to meet me there and I hadn't told him the name on the sign in front. I wanted to save the irony for later.

Ray was impressed with the upgraded ranch property only a few miles north of where he had been searching, and he noticed the name on the sign in front.

"I thought it was a case of smart country boy makes good," he said. "But I start seeing his name on store fronts and I get a different idea about him."

We went inside and found the eating spot on the right side in a separate wing. It was cafeteria style barbeque. We went through a serving line and paid at the end of it. As it was noontime luncheon the place was full, but the service was great and there was no waiting. Taking the trays of food to a table of our own we started eating and discussing our recent strategy. I explained the general lay out the black-top section and how I had driven east on it till it dead-ended at the river near the old Clauson property.

"So he could have easily walked there to a waiting car or hitched a ride," Ray said.

"Yes, but this is his old stomping ground," I added. "He has an interest in new start-ups like Comp-ease Computers."

"You know, Dan, that's right, and he probably knows the FEDS are even protecting him. Why should he go on the run?"

"There are going to be more start ups and that is touchy business. That would be the only thing that might move him out of this area, because he seems to be the only one available with the special nanotech expertise that's needed. He's in business on his own and I don't see him as being in any hurry to spread the word. He can call all the shots."

Ray was still a little shaky on this first day back after the gun-battle. He was doing well but I wanted to be sure that he rested up enough today. I was well pleased with him, but he looked so young. I thought of him as my son, and there was Willie, who was like my father, still in the hospital.

"The way things are we have to hang loose," I said. "We just go on light patrols around here and if we do run across Clauson, we just hang around from a distance without provoking him. We just sort of guard him."

By the time we finished our lunch and had discussed our latest tactics it was the middle of the afternoon. I suggested we move back to the older ranch territory on the gravel roads, where he had been working. We drove both cars back to the spot, where we had parked while we investigated the Clauson ranch. We had the windows rolled down and engines off. I motioned for him to come over to talk a few minutes in my car.

"When we were talking about what sort of thing Clauson might do," I said. "He might have camped right there at the family ranch if we hadn't forced him to leave."

"Yeah, when you think about it, that's about as safe as you can get, a place with secret passages in it. Maybe, a lot of these houses have them."

"He's familiar with this area," I said. "Did you notice any others that might be vacant?"

"Yes, between the one where we had the shoot out and the black-top road. It was close to the river and could have a tunnel that comes out there like the other one."

I told Ray my plan and said it was time for us to go visit Willie. We drove both cars to the hospital and I waited in my car, while Ray made the short visit and returned in a few minutes. We drove both cars over to the motel and picked up a few things. I left my car parked at the motel and we went off again in Ray's car. I had him drive back to the country store that apparently was owned by someone in the Clauson family.

We went inside to a section of the store where they sold sporting goods. A middle aged, bleached blonde woman helped me to select all the things I needed to camp out including a cot and a one man tent. I introduced both of us and the woman said her name was Ruth Smithers. She and her husband had purchased the store from Bob Clauson's grandparents who had moved to San Antonio to retire and live near their son.

"That would be Thad and Wilma Clauson you bought the store from," I said.

"Yes, it was and they were nice to do business with. It was a good deal for us."

"Did you know Bob Clauson?" I asked.

"Yes, he'd often spend the summer with Thad and Wilma, till he got the age to go to college."

We took the cart I had filled with camping items and moved it over to the grocery department. Starting to fill a second cart, I picked up two bags of ice and an ice chest. Then I selected an assortment of pre-packaged food, including a lot of sandwiches and salads. Tossing in several hand sized plastic bottles full of water, we pushed the carts over to the cash register and Ruth rang it all up. I paid in cash and I think she still wanted to talk, because she walked with us to the door.

"Bob used to work at this store for his grandfather, just like his father did," she said.

"Bob must have been a reliable teen-ager saving up for College," I offered.

"Oh, yes he was a very reliable youngster saving up for College. Thad told me he had enough saved to go to the University for the first year. After that, when they found out how smart he was, he was on an academic scholarship the rest of the way through."

Ruth Smithers returned to the store and we loaded up the car. It all fitted into the trunk and I cut open the ice bags and dumped the ice into the light-weight ice chest. I placed all the plastic wrapped food items and the bottled water inside the chest.

I had Ray drive over to the vicinity of the other "safe house" he had spotted. It was vacant and I had him drive on by it. There was a wooded area on the other side of the gravel road and a hilly area that would be good for general observation. I asked him to circumvent the woods and drive around to the next road that would give us access to that same wooded area. When we got there I estimated that it would be about two miles from my drop off point to the hilly area I had designated as good for observation.

"So how long you plan on camping out?" Ray asked.

"I'm playing it by ear," I said. "You can help me get set up and then go back and talk to Willie for a while. I'd like you to try to pass on to him what we've been learning about how powerful Clauson has become. I want him to know how our investigation has slowed down to almost a waiting game."

"So, when do you want me to come back and get you?"

"About eight o'clock in the morning drive by this spot and I'll be waiting."

I took out all the tools we needed to put up the small tent and propped the ice chest on my shoulder like a country boy with a bale of hay. Willie's high powered glasses had a convenient strap I placed over the other shoulder and carried the tools with my left hand, while I balanced the chest with the right hand. Ray brought the tent still in its box.

My survival training area turned out to be more like three or four miles into the thicket. We hiked about two miles and found a suitable place to put up the tent. With Ray's help it only took a few minutes to put it up and place the cot inside. I dropped the ice chest inside the tent also, and I had established my headquarters. Expecting to spend very little time there, I planned to go to more important places.

Ray headed back to his car, and I continued on in the other direction to my designated observation point. In this sort of operation the tent is your house, but you don't go there before you have to. The weather was perfect with a gentle breeze and enough sun for good visibility. When I started hiking up a gradual slope I thought I must be coming to the desired spot at the edge of the woods. I continued to climb and was assured that I was headed right, when I came to the top of the hill and the view I had envisioned. I could circle around this area and observe a great deal even without the field glasses. There were plenty of trees for cover and you could even climb up in one if you needed to.

In this sort of terrain you could not remain unidentified for long in a surveillance vehicle, because you had to drive up too close to the house to do the job. You had to be careful enough doing it the way I was doing it, peaking out from various wooded areas. I started my surveillance. I had to be careful not to be spotted on the skyline. I found a fairly flat place and came to the prone position to use the glasses to scan the entire area.

There were three houses I could see from the gravel road section. I could see about one hundred yards to the left where the gravel road intersected one of the improved roads, there were two more houses staggered on either side of the black-top. Those houses were built on slightly higher ground, looking down on the second safe-house, which I thought must belong to Bob Clauson. I could see flashes of reflected field glasses being used from both of those other houses.

I was convinced that I had been lucky enough to have selected the best location for my surveillance. And I thought Ray must have been right about something going on in the unoccupied safe-house. The FBI was devoted to the protection of Bob Clauson. I had become well aware of that fact, so I thought it must be them watching with field glasses. That assured me that Ray had been correct in establishing that house as a likely place for Clauson to establish a residence. Somehow he had escaped from his original captors and had also been enterprising enough to set himself up with some kind of new guards.

CHAPTER 23

There didn't seem to be any activity at the safe-house for the rest of the day and most of the night. When it was almost dark and I was getting hungry, I made a quick trip to the tent, had a tuna sandwich and attended to anything else to be done before nightfall. Then I went back to my observation post at the top of the hill and remained there till the sun was trying to break out of its resting place and my blinks were getting too short.

Thinking I'd be better off to wait and get a few hours sleep at the motel, I went back to headquarters to satisfy myself that nothing had been disturbed. The tent was zippered up the way I had left it and nothing inside had been tampered with. I had another tuna sandwich, drank some of the bottled water and walked back to my drop off point. Ray was there on time. I hopped in as he was coasting to a stop, in a hurry to tell the good news. "The FBI is watching your safe-house," I said. "You were right about it."

"You were right about leveling with Willie, too. When I told him how things were at a stand still now, he cooled down about trying to get out of the hospital so soon."

I handed the field glasses to Ray. They were too expensive to leave in the tent, and they belonged to Willie not to me. It was good to hear he was going to stay in the hospital where he belonged till the doctor said he could go.

We were out of the troubled triangle and back on Highway 90 in a few minutes. That's the only trouble with Ray. He drives too fast, but nobody is perfect. Of course, he was excited about sharing in any action that might be going down, regarding the safe-house, and he wanted to go back while I slept a few hours. I made the mistake of telling him I hadn't used the cot. He was right though, no need for me to knock myself out.

I had to agree that he could pull a shift watching the house while I slept and we'd be better off. He would come back to the motel and get me at about three o'clock in the afternoon. That gave me time to get cleaned up and to catch up with some sleep. I told him I probably would not stay up that long for my next shift. I was just eager to get a good start on the surveillance and we had done that.

As soon as I sat down on the edge of the bed I was sleepy again and I just crashed for about five hours. I showered, put on clean jeans and a sports shirt, as the uniform of the day. Calling Claire on the cell phone was something I had time to do with my new shift. I caught her at the apartment doing preparation for a case.

"Oh, I'm glad you called," she said. "I had a bad dream, where you were in some kind of a shoot out again."

"We're doing some important surveillance, but no shooting," I said. "What are you doing on your job?"

She was starting a new case in San Antonio and was packed and about ready to leave. I had to let her go. I took out the pocket watch she had given me for my last birthday. She knew I never wore a watch and gave it to me because she thought I might use it. I thought possibly it was meant to stop a bullet some day.

The time was about two forty five PM, so I went outside to wait for Ray. We didn't want to leave our cars in that place to be seen, as it would be a tip off about what we were doing. We were dropping each other off, to avoid having a car parked there to be seen.

When Ray pulled into a parking space near our rooms, I motioned to my car. He parked his and we got in mine to save wear on his car. He had driven my car before and didn't mind driving it back to the motel after dropping me off.

On our way back to the observation location, Ray said he had seen Tom Cobb drive up to the FBI house, which he identified as the one closest to our OP on the top of the hill. There had been no interaction between the two houses as you would expect, if they were all together. By the time we got to the drop off place, we had briefed each other about the things we had observed during our shift.

He coasted down to a momentary stop while I got out and draped the leather strap of the glasses over my shoulder and took off for my shift on the hill. The tent was zipped up and I unzipped it enough to glance around. Things looked the same inside and I zipped it back. I hoofed it up to the top of the hill and assumed the prone position to start my initial scanning of the area. There was no sign of life at the safe-house, but that didn't mean someone wasn't coming and going as they pleased, through a secret passage to the San Marcos River. I remembered how Stark had said they were looking for the boat that had been docked on the river access of the first house. I was sure the FBI wasn't worried about finding it now. For all I knew Clauson could be using the same boat to travel up and down the entire waterway, which was considerable.

There were two cars parked in the driveway of the house closest to me in the improved road location, but there was no activity. The other house in the same black-top section had one car parked in front and it was quiet there also. I didn't see any more reflections from glasses being used at either house. I thought they had learned where Clauson was by now and didn't need to look for him. I was using the glasses more sparingly now and Ray had said he was too. When I saw some movement, but couldn't see clearly enough I would use the glasses, but only when necessary.

A red BMW drove up and parked on the gravel in front of the safe-house. I didn't need the glasses to see that it was Nelda Whittington getting out and walking up the path to the front door. She was more glamorous than I had ever seen her, hair done up like a fashion model's and she wore a stylish blue-skirt and jacket, business suit combo. She was there to see her lover. She had said Bob was like a brother to her, and she had wished that it had been more. I was sure she had been granted her wish.

It presented quite a dilemma for me. If Clauson was in there, I thought he must be as Cobb had thought, he was, a homicidal-maniac. The FBI must have all loaded up in one car and gone to a FEDS conference, McDonalds or whatever. They didn't seem to be around to help Nelda if she needed help. I wondered if Mr. Invisible had gone out through the tunnel to attend to some business and she was sitting alone inside, waiting for him to return. I wouldn't know what was going on unless I left my post and got down there.

Scurrying down the foreword-slope, behind the cover of the trees I made it to the road and crossed it. I kept my body as low as possible, working my way around to the back of the house. Since my last debacle ended with getting two men wounded and a friendly warning from Tom Cobb, I had not carried my handgun. I thought, I better not provide a good target for Clauson, as it would be a one-way shooting match.

From the back of the house I could see a window open. I started from the back corner and worked cautiously to the window till I was close enough to glance inside and see there was no one there. I heard low voices coming from inside at the front of the house. I could just make out what they were saying.

"I'm glad you're at Gladstone now," a man's quivering voice said. "You can do a lot more good there."

"You should be there too," Nelda said. "There isn't even anybody around here, Bob, and you could go with me right now."

"No, we've been through that before, and you know I need to stay here. This is where it all could start. The CIA has discovered people from other countries here in this neighborhood trying to see what they can find out. There are sure to be more. Someone needs to be here to advise the CIA as to how to open up a dialogue with them." "Why does it have to be you?"

"It has to be someone who understands all the sciences that are involved and the consequences of not doing things properly."

"Let me stay here and do that," Nelda said. You can go back to Gladstone and work with the start-ups, openly one on one."

"The start ups are doing fine, I have to keep a close eye on them and I'm doing that." He said. Besides, most of them are

"You're doing it by telephone conferencing?"

"There is someone making calls on them in person to see things in the labs are as they say. If they're not we can shut them down as the contractual arrangement states."

"Who's in charge of the personal contacts with the start-ups?" Nelda asked.

"The CIA is handling it."

"How will I know how to contact you if they move you?"

"I'll call you on your cell phone."

"You're under their bondage," she said. "You were lucky to have got free to call me this time. If you needed to, could you contact the President?"

"Yes, now quit worrying . . ."

"I'll free you now and you can leave with me, then I'll not worry."

"I can't . . . ?

"Well, I'm going now."

The front door opened and closed. There was the sound of the engine turning on and the car driving away. I ran back around to the front of the house and made a break for it across the road and into the forest. In a few more minutes I was perched again on the hilltop. It was completely still except for an occasional bird chirping. There were no cars coming or going. It was quiet at all of the houses. If Clauson had gone he had used the passage-way-tunnel. A white van, probably a rental vehicle, pulled up into the driveway of the house on the east side of the road. A man got out and carried some small furniture items, and a television set from the house to the van.

I glanced at my pocket watch and noticed it was time for Ray to pick me up. I hurried down the hill and back to our pick up point. Ray was there and I jumped inside.

CHAPTER 24

When the white van backed out in the street headed north, Ray was ready to follow it. We traveled a few miles north on it before we got back to civilization and found the streets were marked. The van traveled several more blocks north. Then it turned right and headed east on Travis Street till it joined a main thoroughfare going all the way to the river. We were coming to a decent residential section and the van turned right on the last street before the river named Pecan Avenue. Ray had laid back several car lengths the entire trip, and I had never seen the man before. I thought he had been hired to move a few things for them. I didn't think he was a CIA agent or the FBI.

He stopped his van at one of the houses while we were parked on Travis Street. This was a good spot to observe and not be observed. I saw the van pull in a driveway to the left side of the house and out of sight. In a few minutes he had made his delivery. He drove by us on Travis Street, and back the same way he had come.

I told Ray I wanted to drive by to see the entire lay-out of the house, so he drove past it and parked for a while at the other end of Pecan Avenue. It was an unusual three-story Victorian house with additions, like a cedar privacy fence, which enclosed the entire back part of the house and lawn. The houses without privacy fences had the river cutting through their property at the rear, and some of them had a small boat docked there. I had to assume there would also be some kind of boat or raft docked at the rear of the house we were watching. I thought the CIA must be providing it for Clauson as a guest privilege.

"They don't need any secret passage tunnels here," Ray said. "They're providing the privacy fence, so he can still have access to the river." I noticed that there was a large entrance drive at the very end of the street that went back to a fashionable restaurant called "The Pub" that advertised catfish dinners and libations. It had all of the amenities of an old English pub.

"Pull into that parking lot, Ray," I said. "We can still see the front of the house, and hide our car between other cars."

The break gave time to tell Ray how we lucked out and picked the right safe-house.

We parked there in a good position for surveillance, and Ray bought the coffee. He returned in a few minutes with two steaming plastic cups.

"The lady said to be careful it's very hot," he said. "Anybody else come to the house yet?"

"No, still no takers." It often turns out, if I say a thing is not so, it turns out to be so. I looked up from the coffee and saw two cars drive up to the house. The first car pulled into the driveway keeping to the left and drove up with the front bumper almost touching the privacy fence. The second car pulled in directly behind the other one. They were saving room for at least one other car to park on the right side of the driveway. Nobody got out of the two black late model cars.

"They're waiting for another one, who doesn't want to get blocked in." Ray said.

"I'll bet I know who," I said. "He can't go everywhere from the river."

We were both proved to be right. In about ten more minutes the third car arrived. It was a dark brown car of the same make and model as the other two. They were all full sized Lincolns, but this one came with a uniformed driver. His passenger was Bob Clauson. They all got out of their cars and went into the house together.

"Isn't that Clauson?" Ray asked.

"Yes, it's him," I said. "He gets around; they just had time for a pick up at the safe-house."

"The guy looks more like a movie star than a nanotech research scientist."

"They all circled around him as they went in the house?" I said. "That tells you a lot." I thought his CIA guards had been briefed on the subject of Clauson's importance to the national defense, but somehow they screwed up and let him and his driver get separated from them. I'd bet the next time he goes out they'll all jump into the big brown Lincoln with him. Even then, I thought he might find a way to get away from them.

"How would you guard someone like him?" Ray asked.

'I don't think I'd ever need a job that bad. His nanotech colleagues think he could even make himself invisible if he needed to."

"Those scientists can restructure molecules and atomic structure. You've been keeping up with that haven't you?" The young man was awe stricken about all the possibilities coming into sight with the molecular building blocks. I wanted to put him at ease about it, but I didn't know how.

"All I can say is that you should read articles about nanotechnology. You don't have to be a scientist to get what it's all about. You can get a working knowledge of it by typing it into Google on the internet and reading magazine articles about it."

We sipped the hot coffee and wondered about it all. The lag between technology and culture was growing every day. There wasn't any easy button to push for all the relevant facts to line up and enter your brain by osmosis. You just had to read and discuss things a little each day, and do the best you could.

"Well, Clauson has two ways he can go, the way I see it," Ray said. "He can go by boat on the river or leave the house on foot."

"And either way he will have one of them trying to stay with him. They're at a big disadvantage. He's really the boss and they can't handcuff him to a water pipe. He has to be free to conduct his business.

From what he told Nelda, I gathered that he was still helping Gladstone by working with the new startup companies under their contractual agreements."

"It sounds like he could take his guards with him while he kept up with that, Dan."

"Yes, he might start doing that, when he doesn't need to give them the slip so he can see his girlfriend, Nelda, or have some other kind of clandestine meeting."

"So, what are we supposed to do?"

"We try to be around when he does give them the slip, and find out where he goes and why he does it."

My young partner pushed the ends of his hair out of his eyes and was silent. He seemed to accept the logic in my statement, but he was aware that I hadn't finished the answer to his question. I was still working on it and I had to be honest.

"It's the kind of thing you start into, not really knowing where it's going."

"Yes, I see that, but we do have to know what we're going to do," he said. "Don't we need to arrive at some sort of specific strategy?"

"I thought you might help me with that."

"We would need to anticipate his actions and cover the spots where he might go."

"And what kind of clandestine actions might we be watching out for.?" I asked.

"The kind of actions he doesn't want the FEDS to know about. To meet Nelda again, he would have to make a cell call to her, when none of them are around."

"Before he calls her," I said. "He has to be confident that he can escape, and know the right time for it, so he can tell her where to go and what time to go there.

"Well, let's consider the escape first," Ray said. "Suppose he waits to call her after he's made his escape."

"Yes, we don't need to worry about when he calls her if we can be ready for him at the proper time and place. Where do you think would be the best place for him?"

"At the house, probably early in the morning, when they'd only have one person watching him. They don't want to be so obvious as to have the sentry placed just on the other side of his bedroom door. It would have to be more of a walking and listening post.

We came to an agreement about the general time and place, being early morning in the house. We also thought the only reason for Clauson to escape would be to meet Nelda. The guards were there to serve him, so for anything else he would be glad to have them along with him. It made me go back to my original supposition. His CIA helpers had not been around the safe-house, when I observed the meeting between Nelda and Clauson. I reasoned that it had worked well for him. He must have chosen to go back to them, when he was finished with Nelda. It was as easy as giving them the slip and stepping out on the road to make a cell call. If They had never known where he had gone, so he could use the same place again and I thought he would. I told that to Ray and he thought I was right about it. So we planned to go back to our surveillance spot on the hilltop. I thought I was right on target, but there were a few things that I was wrong about.

Ray said he would go back on the hill, because there were some more details I wanted to check up on. I thought Clauson had someone build the privacy fence recently as an add-on to the house. He must have thought he would be using a boat on the water often enough to be worth it. I was aware that we had barely touched on all the possibilities that were available to him in our brain-storming session. It could be that he didn't have to give anyone the slip, and could just tell them he had some private business. He would probably tell them where he would be and how long it should take. If he didn't show after the time had expired, they'd go looking for him. And even when they were supposed to be giving him his privacy, they probably had the entire area roped off to ensure his safety. I wanted to see what was on the other side of the privacy fence.

When I drove Ray back to the gravel road, we talked all the way about what we had learned regarding nanotechnology and Bob Clauson. The new start-ups had to be successful, before all the miraculous events would start happening, and it all seemed to depend upon this one man. He was responsible for setting up the first molecular, table-top factory. The other new companies, as they developed a small super-laptop, needed help with the final step of the table-top factory. He had sounded like a caring individual, when I overheard his conversation with Nelda and I would have trusted him with the world. That was exactly what I was supposed to be doing.

"There is one thing you should do," Ray said. "Turn off your cell phone, before you get to Clauson's back-yard, because I'm afraid I might get antsy about something on the hill, and call you." He knew I had to snoop around later and he was right.

"I'll do better than that," I turned it off then, as I wanted to be sure not to forget to do it later. The real message Ray was sending was that this was shaky business and we had better both be careful. I thanked him for the reminder as he got out of the car and started up the trail. He looked back for a sign of approval and I gave him the thumbs up.

Driving south on the gravel road to Highway 90, I turned left on it and drove east till I could see the east bank of the San Marcos River. I took another left onto an old two lane road that took me to a place on the river, where you could rent rubber rafts and motors. I parked at a spot where some gravel had been spread around, and I had seen several other cars parked. There was a large metal Shed I assumed to be the place where the business was conducted. I went to the front of it and was met by a youngster that couldn't have been much over twelve years old. I selected the largest raft with the most powerful motor that was available. The young man helped me to put the raft in the water and get the motor started. I headed north on the San Marcos River.

CHAPTER 25

Clauson could have looked out one of his rear windows and seen my approach. Taking out my mini-binoculars, I scanned the houses and found the modern green and white frame. I lowered the glasses to see where the narrow strip of property, to the rear of the house met the water. It was not surprising to see that he had the same type of craft I had just maneuvered up the river. The rubber rafts were probably the most trustworthy boats to be used on the river. Those out for the sport in it might use oars, but you could also mount motors on them if you wanted to speed things up as I had done.

I worked my way along the river bank, walking as close to the water as I could to take advantage of the cover offered by the overgrowth of weeds, bushes and trees. The owners of the properties tended to only clear off the excessive growth, when it had interfered with their own access to the boat docks. In only a few minutes I was hidden behind a clump of bushes and debris examining Clauson's rig, when I saw him coming out of the house with one of his guards. The other man lit up a cigarette as they walked toward me. I took a dive to get a bit farther from the boat. I crawled along low enough to remain out of their view, and closer to a clump of weeds that kept me hidden even if they came to the small dock. They walked on nonchalant and obviously unaware of me, on my knees ready to take another dive.

They stopped just short of the dock to enjoy the view of the water while they talked.

"Be ready to leave about seven in the morning," Clauson said. "I have a blueprint ready to take to the start up in Luling." It was an ideal spot for me to observe and listen to them without them seeing me. Clauson looked even younger now, this close up. The other man tossed his cigarette but, stomped it and lit another one.

"Do you want Steve to drive?" He asked.

"Certainly, that lubberly, limey can do something to earn his keep."

I thought about how sensitive and humanitarian he had sounded with Nelda. It revealed his duplicity. She and Norton had talked about how he was not always careful of what he said and it could make people angry. The other man laughed and nodded his oversized head. He had an oversized belly to match it. He was in a tee-shirt with tan shorts and thin legs that were too weak to support his body.

Clauson was also dressed casually with blue slacks and a paisley shirt. I wondered if he would dress in a suit, when they went out in the morning. Regardless of how he dressed I could see he was the boss. The director had probably selected that crew with the prime criteria being that they not be too smart, so they'd be subservient to Clauson.

"We only need two cars, Bud," he instructed. "We can all get in them easily, and I want you to drive the other one."

"Which one, we leave behind?"

"You may consult the others and come to a consensus of opinion about it, but you should consider which one is in the best mechanical condition. Perhaps you should make the decision yourself. The others could just be confusing, since you're my brain child."

"Yeah, I'll do that," Bud said. If Bud was his brain child, I wondered about the mentality of the others. Then, Clauson turned around, signaling him to go back to the house and they went. I thought sarcasm seemed to be the compensatory mechanism of the day for Clauson. The responsibility of the start-ups that could set off the new industrial revolution was causing a lot of stress. The one in Luling was a simple desktop model. I had seen a photograph of this *personal nano-factory on* an internet site devoted to educating people about nanotechnology. Another name for the same process was molecular manufacturing. It meant that you could duplicate a product as quickly as you would be able to use a computer to copy data files. There was no longer an assembly line needed to put together products, but because products were turned out so fast assembly lines could be used to move them away from the countertop manufacturing unit. It is called an exponential technology, which means that it can reproduce its own means of production.

Clauson had said that he had a blueprint ready to take to Luling. I understood the basic idea of how many everyday products could be produced with a countertop nanofactory, by having it packed with chemical processors, computing, and robotics. The one at Luling must have had those things in it and Clauson was going to supply the blueprint. It was already to go, but I had read there were some real dangers involved, not self-replicating nanobots and gray goo science fiction. You had only to turn on your television set and tune it to the news channels to see how screwed up people were. It was just people behind everything. If a dictator was in charge of the first successful nano-factory, then the world would be under the control of the dictator. And that was the reason we had to be educated about it. I believed the more democratic that a nation was, the more likely it would be to be used to benefit all of mankind, not just a few selfish people who got control of the technology.

I thought about all of this and I could see how the responsibility of using the technology correctly could be causing Bob Clauson to be under a lot of stress. He was human and with the same frailties as everybody else even though he was a genius. At that moment I stopped judging him and started to accept him as I should accept a personal savior of mankind. After all if he wasn't going to save us who else did I have?

Scurrying back to my rubber raft tied to a tree, I collapsed for a few minutes, and was thankful to have escaped from my personal savior, who in a fit of stress might have pumped several rounds into me if he was carrying.

My flashlight became a necessity when I examined the large knot I had tied to secure the raft. I held the light between my upper right arm and my breast bone while I untied the knot. I got in the raft and used the emergency oar to cast off. The motor started easily and I set it on low speed. I juggled the flash for a few moments until I had both the craft and the light under control.

Since I traveled at such a lower speed and remembered many of the land marks I had passed on my way to Clauson's place, I felt a bit more at ease on the return trip. As I came to one location, it seemed familiar. There was an inlet in the river and I pulled into it. The full moon seemed to shine more brightly on the familiar place as I came closer, and I recognized it as the rear of the lot belonging to the safe-house, where I had last seen Nelda. Her visit with Bob was still fresh in my mind, and I thought about the things they said. She had asked him to leave with her and go back to Gladstone. I wondered why he refused to do that. The reason he gave was something about how he would be needed there to discuss the new technology with new arriving members of other nations. He was the nanotech leader of the world, so why couldn't he go back to Gladstone and make the scientists from other nations talk to him in Crystalville. Why did he think he needed to remain in this place out in the boon docks?

I wondered if Clauson's duplicity was showing again, or was it an attempt to cut down on the stress, by conducting as much business as he could in this rural setting. From what I had read about the dangers involved with the start up phase of nanotech manufacturing, I thought he could simply be addressing these dangers. It seemed to me that all of them stemmed from some sort of problem regarding personal relations, or just mistakes being made. The kind of mistakes that resulted in cheapening the molecular product caused danger in the form of health hazards. The cost control people had to loose out to the quality control people. It would be a major health risk to humanity, if the wrong people were in control of the molecular manufacturing. People like me should read about it and learn that only scientists should make decisions regarding the quality of the product.

After engaging in high risk activities by getting into positions, where I could overhear his conversations I had learned things that gave me important leads. I advised my operatives to do the same when they could. Some cases would never be solved unless you became a risk taker. The types of lawmakers we have in Washington have discouraged the federal officers from going in for anything risky these days to avoid scandals. Well, somebody needed to still risk it and I thought it should be me.

So, here I was, putting along in a rubber boat, under a full moon. Reading the water as best I could in the dark, I followed out my inlet and it led to Clauson's safehouse, where we had set up the surveillance point on the top of the hill. Ray would be there now. Since we had never proved this house also had a secret passage tunnel I wanted to find out for sure. I tied up my watercraft and looked around on the west bank of the inlet. In a few minutes I found a place that looked like there had been foot-prints going to a spot by a large cypress tree. I looked around there but found no kind of trap door that could lead into a tunnel. I went back along the edge of the water looking for a small dock. It was much smaller than I would have guessed. On one of those trees that leaned out over the water I found a hook that had been screwed into the tree.

Traipsing around in the moonlight I went back to the spot that looked like foot-prints that someone had tried to cover up by scraping it lightly with a stick. I noticed a hollow spot in the tree. I shined my flashlight around the base of the tree and pushed the leaves out of the way with the front of my shoe. It was great craftsmanship. On the ground on the far side of the hallowed out place in the tree was a hole large enough to put a finger in. I put my index finger in it and pulled up on a stone lid that pulled off like a round sewer lid. There was just enough room for a person to drop into it. If you were obese, better forget it. I could see a small landing a few feet below and three steps leading down to the landing. This was one of the ways that the illustrious leader of nanotech would come and go. We had conducted a search and could only find two of them, but I wondered how many of them he knew about to add to his collection of secret places that he could use for his clandestine meetings. He must have played in them when he was a child. I thought it strange that after he had grown up he would still have such an attachment to those places.

I dropped through the hole and found that I could sit on the landing while placing the lid back in place. There was plenty of cool air to breathe coming from the tunnel and I was through it in a few minutes flashing my light around to look for any evidence left there. There was only the many layered darkness and a few candles placed along the way. I didn't take time to light the candles. There was a pull handle on the door and it was spring loaded. Pulling it slightly I found the springs and hinges had been well oiled.

As I pushed through the passageway door, I was aware of a faint light coming from the front of the house and I switched off my flashlight. An occasional sound could be heard, like the turning of a page in a large book or a ledger. It seemed strange to see an electric refrigerator in a house with an escape tunnel, but there it was in the next room which was the kitchen. The tunnel had been dug and the secret door placed in a short hallway between two large rooms, which were butted up to the rear exterior wall.

CHAPTER 26

The rear hallway with the escape tunnel dead ended into the kitchen, where I had seen the refrigerator and had my first hint that someone was living here. There was also an electric stove, cabinets and all the other usual conveniences. An open doorway on the other side of the room went to a large front corridor with a stairway leading up to the second floor. There was also an open doorway to the left of the kitchen, which was a bedroom with a bed and wardrobe cabinet. Small 40 watt lamps were plugged in from tables or countertops to provide sufficient light and they had all been turned on by somebody.

I picked up a kitchen knife and held it behind me as I stealthily proceeded along the front corridor to a large family room with an archway over the front door. I remembered that the outside of this house had seemed more ornate and larger than the first safe-house where we had the gun battle. There was a plush brown colored sofa chair in my view, as I advanced on the hardwood floor. The sound of another page turning came from the part of the room I couldn't see, and I stepped on a loose board. It squeaked so loud I jumped back.

"Is that you Mr. Brandon?"

I walked on around to the doorway and viewed the rest of the front room. "Yes," I said. "How did you know that?"

"I heard about you from Nelda, and you're the only one that I can imagine coming through that passage at night. Not many people even know where it is."

There was a full length sofa opposite from the plush chair, and the man motioned for me to sit there. "If you can sit a while I'd like to talk to you."

"Thank you," I said. "It's always a pleasure to talk to someone who want's to talk."

I studied the man's face and noticed the unusual wart that extended for about a quarter of an inch between his face and the lobe of his nose on the left side. He was the man it was all about, Bob Clauson. There was no boat tied up outside by the entrance to the tunnel. If he came on the river he would have to have gone by me and I would have seen him.

"You made a joke," he laughed. "It's about how the police try to make you talk."

I remembered the knife I held behind me and I stuck it handle down, into my back pocket. Sitting forward with my hands on my knees I smiled back at him about the cop joke. "Yes, we're always telling people they'd better start talking."

He laughed again with a relaxed, genuine smile. "I appreciate the good police work you've done for Nelda and Gladstone Electronics. I talked to Harold Remkin in a cell conference this morning about our current contracts with the start up companies."

"When I first came on this job, it was to find you and take you back to Gladstone. And now it's supposed to be on some pretense of still looking for you and insuring your safety. Now that we meet, can you clear this up for me?"

"Yes, of course, we're on the same side Mr. Brandon. I'll try to do that. We all want you to stay on the case. I'm going to share as much information as I can with you."

"Well, then, I'd like to know why you shot at us from the other safe house."

"That wasn't me. We know about that incident, and the ones who were involved in it are being investigated by federal officers. It actually was a critical situation, a terrible mistake, and I am really sorry it happened."

"It was a critical situation, or it still is critical."

"In a way it still is critical, but we are in control of it."

"I thought only one person was involved in it. How many were there?"

"Several men had a battle there before you came with the FBI agent. I had been held at another location, by three men and they were moving me to my grandfather's house. They found it was vacant at a more remote location than the other place in Crystalville. They successfully moved me to the river house at Luling, when two other men broke in on us. One of them had a rifle and had been hired to find me and free me."

"How did you get out of there?" I asked.

"With Bud, in his car," he said. "The other man was shooting out the window at my abductors. They also left in a car, and he was caught here by the FBI agent and you know the rest of that story."

"You must have come here by car earlier this evening, Mr. Clauson.

"Just call me, Bob."

"You can call me, Dan. I don't see a car parked in front and I saw you somewhere else not long ago."

"It couldn't have been me, because I've been here going through these ledgers since early this afternoon. And that's all I can tell you, Dan." His voice and manner seemed a bit resentful even though we were on the first name basis.

It seemed as if we were into that duplicity thing again. It could be that he was so good at it and had so much power that I would never be able to catch him at it. He could always hide behind his aloof scientific manner, knowing I could never confront him about his job. To me, it was like he was hiding behind his job. I wondered if he was playing us all against each other in some way.

"I believe what you say," I said. "You had to have been here earlier in the day. I was in this area at that time and I can vouch for it."

"If you were here then you must have seen Nelda from your spot across the road.

"Yes, I did see her."

"And you heard us talking through that rear window. We were upset and it must have been much louder than we thought."

"Yes, that's about it."

"And you were doing your job, and a very good job I must say. It must be very frustrating for you, Dan, and all I can say is just keep it up. These are very uncertain times that we are living in. Eventually things will work out, and it will all have been worth while."

"I hope so," I said. "I've been thinking things will work out, so I'm going to go back through the tunnel to the place I'm calling home and get a night's sleep. Tomorrow all bets will be off and the day will be new. I'll be looking for anything that's not right and how to right it."

"I wouldn't have it any other way," Clauson said.

Wishing I had lit some candles, I made my way with a strong well powered light, but there was no song in my heart, only a bitter feeling that maybe it would all go wrong.

The raft was still there, and I made it back to the main stream. It seemed like a very short time to get back to the place where I rented the boat and I turned it in. I drove my car a little slow and careful from there back to the motel, because I was very tired and I knew that was when a lot of accidents happened. They often happened when people were in bad condition to drive. When I got back to the motel I threw myself on the bed in the room, and slept with my clothes on for about nine hours, before I woke up to the sunshine of the new day. I pulled out my watch and looked at the time. It was five o'clock in the morning. I lay there for a while longer and started to feel a pain in my lower back. I stood up and put my hand in back of me where I felt the slight discomfort. I pulled back my hand and noticed the blood on it. I dropped my pants and had to pull hard to get them down. Then, I saw the kitchen pairing knife still in my back pocket. The blade had cut through my pants, giving me a slight scratch. It was my own fault.

After I got a shower, put on fresh clothes and tended to my scratches I was feeling a lot better. I had some hash browns, bacon and eggs in the motel restaurant and as I was finishing, Ray came in. I ordered another coffee and we had some time to revise our strategy. He hadn't seen anything new from our hilltop OP, and didn't know I had visited the safe-house coming in from the river tunnel. I told him about my personal visit with Bob Clauson and he was just as amazed as I had been.

"It's hard to understand how we could have been trying to figure out when he would escape, so we could observe him." Ray said. "Then he's sitting there in the house like he was waiting for you."

"I don't really know that he was waiting for me," I said. "He told me he guessed it was me, because nobody else he could think of would be coming in through that tunnel at night."

"How did he get around so fast?" Ray asked.

"Many new things are happening," I said. "Maybe he's the invisible man."

I knew I couldn't quit. It would be like quitting the world.

Ray went back on the hill to observe what seemed to be Clauson's favorite house. I got back in my car and drove the short distance from the motel to Comp-ease Computers Laptop Division. I came into town from the south end and got on the new extension of the main street, which was really an avenue with a tree name. Comp-ease was the first company in this new small industrial park. Several other buildings were under construction here and I had to be careful to stay out of the way of the construction crews.

Parking at the rear of the Comp-ease building turned out to be a very bad idea. As I opened the car door to get out I noticed several police cars drive up and park between me and the back dock of the building. A large brown Lincoln was pulling out of the parking lot onto the side street and was being followed by a black Lincoln, when the shooting started. I supposed the first shots were by the police, to try to hit the tires and stop the men. The men in the black car were caught in the parking lot and returned fire while the driver tried to maneuver his car for a speedy exit. The bullets were already hitting my open car door as I dove for the pavement. One of them grazed my head and ricochet into the side of my car above me. Both sides were using large repeating pistols like nine millimeters, or forty five calibers. When you're in the midst of something like that you don't really care who's doing the shooting you just want it to stop. It was about the worst nightmare that I could imagine. It was happening to me in addition to all of the other ridiculously strange incidents I had been enduring on this case. What I had to remember was that this wasn't just a case, it was a war, and I couldn't quit.

How could I have managed to have just arrived perchance, and placed myself in the line of fire of a carload of hoodlums firing back at the police? There was an ambulance or two sounding in the distance and gradually getting closer. I tried to get up and went back down. The next thing I became aware of was a well-muscled-nurse falling on me, telling me to stay in bed that I was in a hospital. I put my hand up to my head and found it covered with several layers of gauze. The pressure I put on it hurt so I pulled my hand away from it and went back to sleep for a while.

When I awakened the nurse was still there and I thought it hadn't been very long since I had the run in with her. She was an attractive woman in spite of her muscles. In fact I was starting to think maybe I had imagined things, or it was a part of my big ongoing dream. I blinked, looked back at her and came to the conclusion that she was real.

"You must have been having some bad dreams," she smiled. "I usually don't have that much trouble keeping a man in bed."

"I'm sure you don't," I said. "I'm sorry my bad dreams caused you trouble."

"If you're up to having a visitor, Captain Torres has been waiting to see you"

My eyes were focusing better and I said for her to call him in. She went out and came back in a few minutes with a tall, dark man with graying side burns. He had set up the road block for me in Luling when

Nelda had been lured there with her research laptop. There was a look of concern on his weather beaten face.

"It's a shame you happened along at that time Mr. Brandon," he said. "I'm glad you're OK. I had two men wounded in that battle and those were enough casualties."

"Anything left of my car?"

"No, it was totaled."

"What was it all about?"

"It started out as a legitimate business appointment. The Compease office had it set up with Bob Clauson a computer engineer. It was after Mr. Clauson had finished his business, that the guy from the other car went in the office and said he was supposed to pick up the research laptop to have it repaired. They refused to give it to him and called us. The first car left and we didn't think there was a connection between it and the other one. When we called for the black Lincoln to stop they refused and that's how it happened."

"Did the office people get a description of the one who said he was there to get the laptop?"

"It was a short, overweight-man with a large head. We have him locked up, but he refuses to talk about it, except to say that those in the second car were alone in the plot to steal the laptop. He says they were hired by the CIA to drive, and to guard Bob Clauson. Of course the CIA claims they have never employed that crew for anything."

"He goes by the name of Bud," I said. "I've seen him with Clauson, and you know the CIA has been known to have people involved with them, doing an agents job. Then it turns out that they weren't really agents, at all."

Captain, Joe Torres was going to search the police computer for a criminal that went by "Bud" and fitted his description. The others in Bud's car were killed. I asked him to also do a search for Clauson's driver, Steve, although he was not in the gunfight. Bud had probably been involved in other business scams, but I had a feeling that Steve would be clean. When Torres had gone, I got up and went into the bathroom and took a look at my eyes in the mirror. There didn't seem to be any sign of concussion and I found my clothes in the adjoining closet. I dressed and managed to unwind enough of the excess of the head bandage wrappings, so I would be able to get on the baseball cap I kept for such occasions.

I checked out of the hospital and called a cab to go to a used car dealership so my insurance company could arrange for me to get a used replacement for my damaged car. In less than an hour I had my car replacement and had transferred all the items from my old office in the car to my newer office in the car. I found my baseball cap and slapped it on my head. Then, I drove to Clauson's latest river house, which was farther north than the other two. It was a modern structure, not a safe-house with an escape tunnel. The first safe house, where we had the shoot out had originally belonged to Clauson's grandfather. It was the one furthest south with the escape tunnel to the river. I thought Clauson must have been there and escaped even though he said he wasn't there. It must have been the place where he was taken after he had been kidnapped in Crystalville. According to my theory Junior and his cohorts must have learned about the safe house at Luling from him. It had always been a convenient arrangement for someone who wanted to remain incognito like Clauson. He was still using the next one south of it now and it made sense to me that it was a good place for him to be living while he worked with the people at Comp-ease.

CHAPTER 27

The brown Lincoln was parked in the garage of the modern, green-and-white-frame river house. There were no other cars to block it now, so why not park in the garage? I thought someone must have advised every law officer, in the same way that Tom Cobb had advised me, regarding the status of Bob Clauson, world-class, nanotech champion of the world. No reason to hide out from the law for him. There was no evidence that would link him to Bud's criminal venture.

In such an incident, involving an ordinary citizen, I'd have mentioned the conversation I had overheard between Bud and Clauson. Then, they would have been brought in for questioning. But I could see no reason for Clauson to be involved in an ill-conceived, dim-witted stunt to steal a research laptop. He had just documented himself with an official visit as a special nanotech engineer, and we all bought the idea that he was not connected to the carload of criminals. That they all lived in the same house and drove around together was attributed to the government or some of the government's bad choices in the way of picking guards for him. If you looked at it another way, you might say it was the right choice. The "outfit" produces the most feared guards, of all. If they get caught by the law, the CIA doesn't get the heat from it. It was the perfect ruse.

I rang the doorbell on the front door, and waited. Steve, the driver of the brown Lincoln was out of uniform, when he opened the door. He was dressed in tan khaki pants and a red safari shirt. I asked him if he had the rest of the day off after the big shoot out and he didn't seem amused. He was cordial and invited me into a well furnished entrance parlor. We sat in leather covered chairs to talk. He already knew my name and all about me. Speaking with a slight British accent, he was smart and well educated. He introduced himself as Steve Wingate.

"Yes, I think we should ponder over this terrible thing," Steve said. "I can't say I'm surprised that they would do something like that. They were like an accident waiting to happen."

I thought the sensitive, red-headed man with freckles was quite a contrast to Clauson's other helpers. "They were supposed to be protection," I said. "And you're a professional driver and a bit more."

"I'm more, I hope to God, a university trained professor of cybernetics and a cyborg with a computer chip implanted in my brain."

Cybernetics, I had learned, was a study of control and communication between people and computers. If he was a cyborg, it meant he was a combination of man and machine. The implantation of computer chips into his body resulted in the ability to think in more dimensions, with improved ability to analyze and compare. He would also have a great enhancement of his memory. All of those things are a great help to the learning process. He would make a great assistant to the world's number one nanotechnologist. He was hired by the CIA Director to find Bob and free him. When he came into the picture, he said he was desperate, and Bob was being held in his grandfather's safe-house. It was his decision to hire Bud and the other two men to break into the safe-house to free Bob Clauson. Of course, there was more to it than that but he said that was all he could tell me at that time.

"Is Bob getting a chance to rest up, some?" I asked.

"Yes, he was fretting around here so much that I told him to take half his sleeping pill and go lie down, and he did it."

"Does he make a practice of sarcastic humor or is it negative thinking?"

"Oh, it's humor, a way to endure all the stress. He's the most positive person I've ever known, and he's got to be that way in his present capacity to call the shots for global nanotechnology.

"What's going to happen about the protection factor now?"

"We get some real CIA people this time," He said. "You can see our vulnerability until they come. It was a good time for you to come to us, Mr. Brandon. Can you stay till they get here?"

I said I would stay, and felt some of the stress myself. Taking a few deep breaths did the trick for me. As hot shot PI you come and go as you please and things usually turn out for the best. But it was a surprise to be met by such an erudite scholar, accepted unconditionally, and asked to join the security team. If I believed what I had been hearing and reading about the present geo-political situation and the importance of nanotechnology, this was a crucial time for the world. It was also important for America to be in a leadership role, when the first molecular manufacturing started full scale. If a dictator had control, the people of the world who needed life giving things, like water, might not get it. Dictators used killing fields to keep people in a state of fear, so they could stay in control. They had not won awards for their humanitarian endeavors.

Steve Wingate was not only surprising, but also a challenging influence on me. I was already checking the windows and doors, looking around the neighborhood. There was a security system control pad just inside the front door. He caught my drift and went in the other room for a few moments and returned with an aging German Luger, and placed it on the end table next to my leather chair. It was WW II vintage, and a well-preserved artifact of the German war machine. The Brits always seemed to have respect for anything German. I thought their antagonism had produced something spiritual between them and could understand how Wingate had found it desirable.

"You would know how to use this better than I," he said. "I'm adequate in the martial arts, but not much with the artillery."

"I haven't been carrying a handgun lately," I said. "How did you know that?"

"It was a guess, you see. I noticed others had put their guns aside, when they were around us. I thought maybe you had decided not to carry, to be sure you didn't shoot us."

He had read my mind. I was afraid of shooting them. I picked up his handgun, secured the barrel under my belt, and went for a careful trip around the grounds. Examining every inch of the property along the river bank in back of the house, I found nothing unusual. There were no marks in the muddy area by the boat dock that would indicate fresh shoe or boot prints, and the boat rig appeared to be just as I had seen it last. I even checked around the weedy spot where I had lurked while doing my own spying, and found no debris of any kind that hadn't been there before. Removing the mini binoculars from my pocket, I used them to scan the adjoining property on both sides of us. There was nobody around except a few children playing in their own yards. I trained the glasses on the land on the other side of the San Marcos River, and that was when I noticed a circle of parked vehicles and several people looking back at me with their own binoculars.

Steve Wingate met me at the door, when I returned to the house. He was talking on his cell phone and looked up to say, "They're already here." He seemed relieved.

"You mean our protection has arrived, I assume."

"Yes, that was the Director." He closed the phone, and put it in his pocket.

"Did you get any of the particulars?" I asked.

"He said for me to inform Bob of their policy. He said Bob will have to give them his plans for every move from now on, so they have time to block off the areas and keep us in protective locations at all times. It's just what Bob did not want, but the Director said it was a necessity from now on."

"You've met the Director? I asked.

"Yes, when I was working with the British Intelligence to set up a program to implant chips in agents very much like it was in the James Bond movie. The American CIA Director was visiting there, and I was introduced to him. He eventually recommended me to go to America and become Bob Clauson's assistant. Incidentally I tried to convince Bob to accept a more controlled environment.

"I have a feeling he's going to be ready to accept it now," I said. "Are you going to wake him up soon?"

Wingate did go to check on Bob, but came back in a few minutes without him. Bob was still sleeping so well he didn't want to bother him yet.

"There are three agents on their way here now, and they can be arriving at any moment. Do you want to be here?"

Deciding I'd be better off without three more agents, to possibly impede my progress on this case, I said no. I thought there were already enough of them in my life. Since this was a lot more than a case, and I couldn't quit, it was all the more reason for me to do the best I could. I needed to continue to use my own methods. Steve Wingate seemed to agree with me, and wanted to consider me as another possible resource and we exchanged cards and cell phone numbers. Instead of putting his card in my glove compartment with the others, I put it in a special place in my wallet.

I went back to the new version of my old car. It was gray, the same color as my old car and somewhat nondescript in other ways too. When I got inside and it started right up, I decided I liked it as well as the other one. It was the same year and model, ran great and I didn't have to pay anything for it.

Driving to the end of Pecan Street as I had done before, I pulled off into the restaurant parking lot and into a space, where I could observe the front of Clauson's house. In a few minutes two cars pulled into his driveway. A young blonde woman got out of the first car, and two forty something, dark complexioned men jumped out of the other one. Steve opened the door and they all went inside to make their introductions. And now, I'd recognize the agents, which would be good in an emergency situation.

Since I was already in The Pub's parking lot I opted for an early lunch, and went inside to try the catfish dinner. It came with fries, two vegetables and all the tea or coffee you could drink. The chicken-fried steak also looked good and I thought I might try it the next time. The check arrived and I put down enough to cover it and the tip. As I started to get up my cell phone sounded off and I answered it. Ray had just learned that Willie was getting out of the hospital and was going to start relieving him at the surveillance point. It was good to know Willie had mended well, and I wanted to continue that surveillance. After I brought Ray up to date I gave him some instructions.

"If anyone goes in that house I want you to call me and tell me about it," I said. "I'd like you to do as I did, then. Work your way down the hill to the rear of the house and if Clauson is in that house I want you to stay there and call me right back to tell me."

"Do you think he's some kind of double agent?" Ray asked.

"No, but I'm not sure about it. It would be bad for this country if he was. There are things about that house and the other one where someone shot at us, that we need to find out."

"I'll pass all this on to Willie, and if he's on the OP and he knows Clauson's there with someone, should I tell him to call both of us?"

It was a good question, and I said yes, it would be a good idea and we closed our cells. I didn't want to alarm him, so I didn't mention the possibility of making an arrest, but I knew he was thinking about it.

CHAPTER 28

It seemed that things were going to be handled better with Clauson, now that the hoodlums were out of the picture, but only time would tell. I couldn't imagine him giving Steve Wingate the slip, when he met Nelda at the safe-house we had under surveillance. I thought that Bob and Steve were very close, as Steve was a cyborg and an expert in cybernetics. Now, I was aware of Bob's sense of humor. Since I knew about Steve's true attributes, I could appreciate the joke about Steve earning his keep by driving Bob's Lincoln. It was not a put down, and I could see Steve laughing at it himself.

Now that I thought about it from a different frame of reference, I wondered if Steve was aware of the tryst between Bob and Nelda. He could have even been the one who had driven Bob to the house to meet her that day. Our surveillance had not been at its best at that time. In fact, Steve could have dropped Bob off at the spot in the back of the house, where the road jogged off a bit to extend into a place where the inlet of the river curved away from the property. Bob could have walked from there to the rear door without being seen from our OP on the hill. He might have gone in the back door only a short while before I arrived at the rear of the house myself. They had not been getting along too well together. She had started the argument, by suggesting that Bob get out of there at once and leave with her. He refused and I could tell she was quite upset, when she left alone. This was another question I needed answered. Why didn't Bob return to Gladstone with Nelda? That was what they all seemed to have wanted. In my last meeting with Harold Remkin he had asked me to bring Clauson back. Things had changed a lot since then, and I assumed that they all had got used to Bob doing things the way he thought best. And I had realized that I had a new, greater, overall, boss, which was serving my country at a time when it really needed me. If I failed on this job it could literally be the end of everything.

Driving back to downtown Luling, where the jail was located took only about twenty minutes. Joe Torres was quite concerned about his two men that had been wounded in that gun battle and he wouldn't forget it. He had the prosecuting attorney ready to ask for a maximum sentence for Bud. They felt he had been responsible for what had happened whether or not he had fired any shots himself. On the other hand Bud could know who was in the safe-house with Bob and had fired those shots at us. It was probably the same person who had killed the three men who had abducted Bob. I didn't think it was Bob. He told me himself that it was someone else and I believed him. I was forced to confide everything I had learned about our countries involvement in nanotechnology and the important role that Bob Clauson played in our national security. It was not easy. I talked to Joe for over an hour, before my story started to ring true to him and he admitted that the overall prospects for this country could overshoot his own local concern.

"Exactly, what is it you want me to do?" Joe asked."

"I'd like you to talk to this guy, Bud. Let him know the prosecutions plans for him and how he can do a little better if he will tell us all he knows, about what went on in that safe-house. I need to eliminate the possible option that Clauson did the shooting."

"What if Bud refuses to cooperate, or says he doesn't know who the deer-rifle shooter was?"

"If you and the prosecutor's office strike out then, let me talk to him."

"I can tell you more about him now that we got info from the computer. His full name is Bud Edgar Wilson. He was involved in scams against home owners. He was little more than an aluminum siding man before he got in the gang. The two guys who got shot to death were big time shooters for the mob a while back. Either one of them could have been behind the sights of that deer-rifle, thinking that was the way to protect, Bob Clauson."

Joe promised to call my cell number as soon as he got any kind of answer from Bud Wilson, telling what he knew regarding the shooting incident at the safe-house where Ray and Willie had been wounded. It usually took a few days before they had the scare built up enough for someone like Bud, so that he would decide to cooperate to get his sentence reduced. The police were also eager to solve the three murders committed by the deer-rifle killer. The time was right and I struck while the iron was hot. It was a priority, now, to find the murder weapon, and Joe Torres assigned two of his men to accompany me to the river-house on the San Marcos River. He would have a search warrant if we needed it.

"I don't think we need a search warrant," I said. "They're going to be cooperative."

"What else do you need?"

"If you can supply four or five plain clothes people to search, it would be great, because there are a lot of places to hide things there. The occupants of the house are the good guys, now, and the sooner we finish the search and get out of there, the more we will be establishing a good future relationship with them."

He was a good listener, and I took advantage of his patience to give him all of the reasons, why, it was important to treat this group with a maximum of diplomacy. Joe was quick to agree with me, and there were three unmarked cars full of plain-clothes police officers ready to go before I made my cell pone call to Steve Wingate. Steve answered his phone and I made my well-thought-out proposal after we had indulged in the necessary verbal social amenities.

"Oh, Bob has done well with the new people, you'll be glad to hear."

"Yes, that's great," I said. "I've been thinking about his welfare."

"Is there something in particular you're thinking about?"

I said yes that there was something very important to the promotion of Bob Clauson's public relations. Laying the groundwork for my request to search the grounds, I told him about all of the evidence the police had accrued against Bud Wilson and the two deceased members of the old guard, who had fortunately been expelled from the team. He was overjoyed to hear me say that if we found the rifle they had used at the safe-house, we would be sure to prove one of those three had done the shooting and not Bob Clauson. His cyborg mentality seemed to be ahead of me, and in fact, he suggested I have the police do a thorough search of the grounds at their river-house. It was a classic example of how criminals often held onto a murder weapon too long, because they thought there might be one more time to use it. With three previous murders to its credit you would not want to hide it in the house where it could be found more easily.

"We won't bother you inside, to search there, except as a last resort," I said.

"In that case we'll give you time to get out first, if you like."

"Very good, Brandon, come ahead."

With the police sirens blasting out, we were there in minutes. I drove my own car, but followed in their wake, so I got there at about the same time. With all of his cyborg confidence, Steve Wingate was waiting at the door, and came outside to shake hands with me before we started. He was obviously impressed with the magnitude of the search provided by the small town, as I was. I suspected that Joe may have been forced to dip into the ranks of the local volunteers, but I raised no question about it. I thought a beggar like me could never be a chooser.

Most of the locals were clad in jeans, sport shirts, or blouses. They had been told what they were looking for. They spread out from the back of the house and proceeded toward the river like they knew what they were doing. I worked along only a short distance ahead and went directly to the boat dock, thinking I would make an early discovery and save the others a lot of time and energy. It wasn't that easy.

In the first hour we had gone to the river and back several times and seemed to be slowing down each time, for a more careful look. On the first trip down to the river bank we had removed the motor and taken the boat out of the water to examine it. We tramped along the edge of the water, and one man stretched out to look under the short dock. He fell in the water and when I grabbed his hand to pull him out he almost pulled me in before several others came to help us. The man laughed and said he got a very good look under the deck and there was nothing but mud under there. Scraping some of the thick, murky-brown deposits from his pants with a stick, he passed the stick over to me and I scraped some of it off my shirt. The rifle could have been attached to the underside of one of the planks of the dock, so we went into the second hour with quite a few chuckles, and the positive reinforcement that, at least, we had accomplished something.

As we returned to the rear wall of the house I noticed a lot of attention was focused on the roof hangover and other things we hadn't paid much heed to before. The privacy fence became a new attraction for us, and I opened the gate on the left side and advanced into the new space at the side of the house. This was the side I wasn't familiar with. All I could remember about it was the huge oak tree, which cast a shadow all the way from the fence to the front sidewalk on Pecan Street. It made an ideal car port. In fact, some pea-gravel had been strewn around by the trunk of the tree and a car was parked there now. It was the other black Lincoln, the one they had left behind on that senseless, last fatal event. The window on the driver's side was open and the keys were still in the ignition. I pulled them out and opened the trunk with the remote which was on the key ring next to the ignition key. I pointed it at the trunk and pushed the button. The lid popped open and there it was, wedged in behind the spare tire, a lightweight Remington 308 with a full box of shells. I left it where it was, and closed the trunk. I called Tom Cobb's cell number and when he answered I told him what we had. He said he'd meet me at the LPD. Steve Wingate stepped outside for a moment, when he heard our commotion, and I told him the good news before we started back to Luling.

I thought this was about what you might expect, with the kind of shape we were in these days. Here was the great hope of mankind being guarded by those people. I sure wasn't going to be broadcasting this story and I didn't think any of the others would either. By the time the Lincoln was gone over by the LPD and the FBI we should have the evidence we needed. I was ninety nine per cent certain that neither Bob nor Steve had ever touched that rifle.

When I drove the Lincoln into the LPD compound Joe Torres was waiting for me. He motioned for me to get out and one of his men drove the Lincoln through an open overhead-door into a large work area, where I could see agent Cobb and his team of experts were ready to get to work along with the LPD. These extraordinary circumstances provided an extra boost of motivation to all of us involved with law enforcement to do our jobs better and quicker than we ordinarily would have. I was keyed up as we went into Joe's small office. He was a large dark man sitting behind his desk. He punched the telephone console so we wouldn't be disturbed.

"We've got double good news today," Joe said.

"Good, I'm used to getting double bad news."

"When I told Bud that you found the deer-rifle, he caved and spilled his guts."

"That was well timed," I said.

"Yes, it was," he laughed. "Bud was there through it all, and told all. In his statement he fingers Ted Schmidt as the shooter. He was hired to get the guys responsible for the Gladstone laptop theft and for what they did to Bob Clauson."

"Who hired him?" I asked.

"Bud claims it was a guy Schmidt met in a bar in Crystalville. Schmidt told Bud that the guy paid him six thousand dollars to do Campos, Gilbert and Oliver. The guy gave Schmidt an extra Four thousand to help Clauson, whatever that meant. I thought you might like a copy of Schmidt's mug shot." He handed over the photo and I dropped it into my pocket. I did think I would have a use for it.

"Did he get any description of the man who was supposed to have hired Schmidt?"

"No, but Bud said he couldn't ID him, as he didn't think he'd ever seen him."

"I'd bet my bonnet Bud has seen him with Schmidt," I said. "He'd fear him."

"Yes, Bud said the guy also provided the deer-rifle, so I'm going to keep plugging on that before we make any deal with Bud."

CHAPTER 29

The good news I had received from Joe Torres sent me into the heavy traffic on Highway 183 without giving it a second thought. I kicked up the speed and was soon cruising along north of Luling on the express highway to Austin. I took a cut off that put me on north LH-1 and I was in Crystalville in a few minutes. I parked in back and went into the detective's office. Charlie Sloan was alone at his desk typing something into the computer. He looked up at me with his permanent wise guy smile and set aside his paperwork.

"You look like you might be onto something," he said. "What can we do you for?"

"I need to find out if anything came out of checking the neighborhood near Marsha Brooks' garage."

"We did get one man to admit he saw an older man break the window and go in through the window. We brought in several men, but he couldn't identify any of them."

"I think I know who broke in to get the rifle." I said. "I need you to get the neighborhood man ready to identify him as soon as possible."

"He works at night and sleeps during the day. He may be up by now."

"How about this," I said. "My man lives close to here and I'll go there and call you back. Then you can tell me if your man's available and you can bring him to the office of the Crystal Palace Hotel."

We had it set up between us, so I went back to my car and drove around to the front of the CPD building. I turned east on Lake Crystal Drive, driving onto the loop that took me around to the Crystal Palace at the LH-1 access road. I parked in front of the hotel and got out. There wasn't anyone at the front desk and I went on behind the desk and looked down the short corridor where Albert seemed to spend most of his time. He must have heard me moving around in the outer office and stuck his head out of his room.

"Will you be around here for a while, Albert," I said.

"Yes, the rest of the day." He rolled his eyes and ducked back in his private place. I'd be back to see him later, when I'd have more time to talk. I went back to the car and called Charlie on the cell. He told me his man was up and ready to go. They would be here in about ten minutes. It didn't even seem that long before Charlie drove up and parked next to me. We all got out and went into the front office.

Albert was at the front desk and our neighborhood man spoke out at once.

"That's him," he said. "He's the one I saw break into the garage." He only glanced at Albert as he spoke. I thought it was quite likely he identified Albert as the guilty party just to get this business over with and off his back. It wouldn't be the first time somebody had done it. He would figure not much would happen to Albert anyway.

"Why are you making this accusation?" Albert asked.

"Because it's true, and I want the police to leave me alone and be done with it."

Since I wanted to get at the truth, I was pleased with this natural progression of the dialog. It was noteworthy that Albert didn't emphatically deny it. Charlie was assisting me in the application of the silent treatment, and the next thing we knew Albert was inviting us all to go back to his reception area, have a coffee or soft drink and talk things over. By the time we got back there, and we all refused to have anything to eat or drink he broke down and confessed that he had been involved

"Look, I was in an embarrassing spot, all I was doing was taking back my own property."

"There's a little more to it than that, Albert," I said. "It's about what happened to your property after you took it back. Can you explain that to me and Detective Sloan?"

"Yes, the incident at Marsha's garage was necessary. I couldn't ask her for it back."

I suggested that Charlie drive his man home, and then come back. He did that and while we waited, Albert and I sat down and had a cup of coffee.

"I swear to God Dan," He said "I didn't shoot anyone with that rifle"

"I want us to wait for Sloan to get back," I said. "Don't say any more till then."

He sipped his coffee with the mournful look on his face that remained there for as long as I knew him. I thought Albert had been sorry since the day he was born.

When Charlie returned, he poured himself some coffee and assumed a mood that was sour for him. I motioned him back to the chairs and he joined us.

"Albert said he didn't shoot anyone and I asked him to hold off till you got back."

"I thought, since the two of you know each other, you might want to start off, Dan."

Charlie was being considerate, and that made him the good cop. I would take on the guise of the bad cop if it seemed advisable to use that age-old procedure. However, I thought it might be better in Albert's case to just continue to give him more rope.

"You must have had important plans for that rifle," I said. "Would you tell us what you did with it?"

"Well, there was a man who came to our restaurant-bar, and he asked for me. He had heard I had an interest in Gladstone, and he said they were in trouble, but he could help things."

"Was his name Ted Schmidt?" I asked.

"I don't think he used that name."

"He probably wouldn't have used his own name." I handed him Schmidt's photo. "Is that the man?"

"Yes, that's him."

"So, what did he say to you about helping Gladstone?"

"He said he knew three men, who had overpowered Bob Clauson and were holding him and they also had the super-laptop. He said to release Bob he'd need a rifle with a scope. That was when I got my rifle back and gave it to him."

"Was that all he needed?"

"No, I had to give him some money too. Then, when Bob and his laptop were free I paid him some more money."

"How much money are we talking about?" I asked. "And specifically what was Schmidt supposed to do?"

"I'm not stupid," he said. "I had a witness to it all, and he also recorded the conversation. We assumed that Schmidt was in on the kidnapping and it was like paying ransom money. The witness was my lawyer, and we made it clear that the money was to pay for the release of Bob Clauson, without killing anyone."

"If you had let us handle that for you, you'd have saved that money and the legal fee, whatever that amounted to," Charlie said. "It didn't turn out that Schmidt handled it that way. He murdered three men and wounded two others."

"As I said, I didn't think any of that would have happened and I can prove it," Albert said. "My lawyer has one copy of the DVD audio-visual we made, and I have one of my own."

Suggesting that we observe this DVD, I sat back and sipped my coffee while Albert got computer equipment together, so we could learn about the evidence. The equipment was high quality, and the conversation proved to be just what he had said it was. It specifically referred to the money as ransom money to secure the release of Bob Clauson. Evidently, Schmidt didn't care what they called it as long as he got the money. It mentioned the additional four thousand dollars to be for the protection of Bob Clauson. Apparently Schmidt was double dipping on that one, since he was already supposed to be protecting

Clauson. The entire payment of ten thousand dollars was paid in advance, just like a ransom would be paid. The story told by Bud Wilson was quite different, but Charlie and I agreed that a jury would not believe a convicted felon's story over Albert Hoffman's carefully documented version. It was all wrong, and we made sure that Hoffman understood that we didn't approve of his way of doing business, before we left there.

"You still aren't off the hook, Albert," I said. "You still could do jail time as an accomplice to murder for providing the weapon." We left him standing there with that same hang-dog expression and went back to our cars. I knew Charlie would type his report on Hoffman, and the CPD would be contacting Albert's lawyer to request the DVD compact disk, and any other information that office had regarding Albert Hoffman and his transaction. This was only the beginning. They would need more evidence than what was provided by this preliminary interview, before they took Albert into custody.

"I thought I'd sleep in my car, before I'd use the room provided to me by the Crystal Palace Hotel. There was no telling what good old Albert could be involved in before this case was over. I no longer trusted him and didn't want to have any further obligation to him. Smoldering inside of me were the disgusting results of several incidents I had endured with the people at Gladstone Electronics, and I didn't feel impelled to drop in there to make a progress report. It was dark by then anyway.

Deciding to spend the night with someone I trusted, I drove the short loop to the access road of LH-1 and drove a few blocks north to the traffic light where I turned around to South LH-1 and drove to Austin. I owned two door keys to two separate residences in Austin. As I sped along in my own night-time version of driver's hypnosis, my mind was reviewing, at random, the events of the past few days.

One thing I was sure of had to do with relativity. Everything is on a continuum and no matter how hard I tried to resolve the events in my life I would never be able to completely resolve them. It would just be moving things around. When an event had been resolved as well as I could do it, I would relegate it to the vast ranks of the unconscious mind and that meant forget about it. That was what we were doing during our entire lifetime. We were just making selections about what things were important enough to remain in the most conscious parts of our mind. I think of this as the control center. Sometimes it's hard to keep things going right in this part of your mind.

My control center had been really screwed around with lately. Because of the exigencies of the situations, I had been forced to move things around up there too fast. As I'd try to shove things over to the unconscious side, some of the things kept popping back up to demand my attention.

Things like cyborgs and silicon microchip implants that made you a lot smarter than you really were, and things regarding ethics kept popping up. I thought an intelligent chip implanted into the brain of a person with Parkinson's disease was a good idea. In that case tremors on the way would be signaled to the brain and the brain would stop them. A person with a hopeless disease, that keeps them stumbling and falling is able to walk and regain some hope in their life because of the implant. For me, there is a big difference between the later case and someone who gets a microchip implant to be smarter than they really are. Here is where I have an ethical consideration about it. If a person was having a hard time keeping a job to support their family maybe they needed the chip implant to make them smarter, but where do you draw the line. I wondered if we were headed for a time where everyone had to have it. If you didn't have it would that make you a societal outcast? People might be standing in lines to get the implant, the way we stand in lines today to get a flu shot. This case had me so confused, I thought I might be one of those people who needed the brain chip implant to be successful on my job.

CHAPTER 30

As I went speeding along on the freeway the black shadows that surrounded my car seemed to be getting lighter, moving into the gray area. I wondered if it was just my imagination. Things seem to be largely in your mind. What if an employer could give employees a vacation silicon-chip-implant? We had the technology in place now to do it. There would be different ones, like Hawaii, Mexico, Italy, Germany, and The Grand Canyon. You could sign up for whichever one you wanted. Now, what if each of these vacation implants would be standard seven day trips that would take place in your mind in one hour. I wondered if some employers might say it was an equivalent to having seven days off with pay. That would give the employees the opportunity to have their cake and eat it also. They could go right back to work and go on making money after the employer had given them a free vacation with pay. Some employees might think it was a good deal. It would probably depend a lot upon the quality of the microchip implant, and their personal feelings.

If Claire, was back from her business trip I might consider her as an equivalent to a seven day vacation with pay. There wouldn't be any kind of seven day vacation that I would accept in the form of a one hour free implant, unless she could go along too. I'd rather have some time off without pay, and plan my own vacation in real time.

Thank God, I was coming into the Austin city limits. I had too much time to think on the road. You could get lost in mind games, and I think I was loosing it. I pulled off at the Caesar Chavez exit, and took it to the anomalous downtown apartment complex, where Claire had her own place to hang out if she was in town. There was a parking space open near her ground floor apartment. I parked in it and went inside. Her Land Rover had not been parked in its spot, but I was thinking positive. I thought she should be finished with the case by now, and just be out to the store for something. Now that I thought about it I often had insights in a case that solved it, while thinking on the road. We shouldn't knock it.

Taking out my door key, I made a quick stab for the key slot and it went in. Pushing the door open quickly, I stepped inside wishing there would be no note, but there was. It repeated the information she had given me on the phone, and included a phone number in the San Antonio Court House building. There was also the name and address of the Hotel where the accommodations had been made. It was past midnight or I would have driven to San Antonio. The note said the usual things like what was in the refrigerator, but there was something new. It said if I came in late and was really bushed and keyed up I might help myself to an ambien from the medicine cabinet. I had never known Claire to use any sleeping pills and that one was a strong one. I thought all of the research she had been doing for court cases, must have been causing the sleeping problem. I was concerned about it, as I had seen some TV News reporters telling stories about how some people had taken that pill, got up to soon and went driving around in a daze that caused them to have a wreck. I was thinking I would try to reach her at her cell phone number, and my own cell phone sounded off. It was Claire.

"I thought you might get back to Austin, before me and be disappointed," she said. "And I wanted to tell you I'll be going back tomorrow morning."

"I'm at the apartment right now and I am disappointed, but that's not your fault."

"I know it's not. It makes me feel bad anyway. I'd leave now, but there's a meeting in the morning I have to attend to wind things up. I'd be missed if I wasn't there."

"Sure, I can understand that," I said. "There is some business here to attend to tomorrow, so I'll see you then, maybe, for lunch."

"I should be back there by eleven thirty or twelve, so is that a date?"

"It certainly is. I'll be at the apartment first. Say, be sure you're well wakened before you start back, if you take that sleeping pill."

"Oh, yeah, I only tried it one time, but I haven't been taking it lately. Don't worry about that."

Telling her I wouldn't worry about it, but at the same time worrying that she might go back to them still left me in a dilemma. I wondered what she would have thought if she had known that I had almost called her about it before she called me. She'd have thought I was bonkers and maybe I was. I kept my mouth shut about it, just made pleasant cooing sounds, and she did the same for a while, until we closed our cell phones.

I took a shower and used my electric razor, so I'd be ready to go in the morning as soon as I had my coffee and a bagel. She had taken it very well, when I mentioned the sleeping pill, knowing exactly what I had heard about it to worry me. It made me feel so guilty of my lack of confidence in her, that I went to the medicine cabinet and popped one of the danged things.

Sleeping a bit better and longer than usual, I went out the door at nine o'clock with a bagel stuck in my mouth and the coffee cup in my hand. There was no problem with my driving and I was not in a daze. In fact, I felt up to the great task of a cross-check on two governmental agencies. I proceeded as planned to the Federal Building, only a few blocks away from Clair's pad. It was a respectable hour for doing business and I didn't have any appointment to worry about.

There was a parking space open in the same block as my destination, so I put quarters in the meter and entered the magnificent work of architecture that our tax dollars had paid for. The same poorly-dressed, elderly lady who was always present upon my entrance asked me what I was doing there. I answered that I was supposed to see agent Cobb like I had an appointment. She smiled as if she was well aware that I was full of it. I hastened to say, "Agent Stark would be fine."

"Come this way," she said, motioning down the corridor. "Go to the second office on the right." I said thank you, and went. He was there just as she had suggested, typing some kind of FBI report into a desktop computer.

"We got some news for you, gumshoe," Stark said. "Those shells at the old safe-house matched the shells used to kill the two riding the motorcycle and they matched the ones that were used to kill Tom Oliver. Of course, you know we've got the rifle that was used and the one who used it, a man by the name of Schmidt."

"I've been worried about the security of our nanotech man, Bob Clauson." I said.

"Oh, time to cross-check us eh?"

"You have to know about the big screw up of contracting out the guards."

"It's all over now and it's best to forget it, gumshoe."

"Well, it wasn't an FBI screw up," I said.

"And you think I'd comment on another agencies screw up."

"No, I guess not, but I can tell you some things about it," I said. "Clauson has an assistant, who's a cyborg with a brain-chip transplant, and the new guards are supposed to be legitimate CIA people, but how do we know?"

"Like I said before, Brandon, Clauson's going to be fine."

"And does he check out to be what they say, our nation's first line of defense?"

"Of course he does, he's controlling that nanotechnology business for us. If the wrong people get control of it they'll have the biggest nuclear weapon there ever was. You know we can't have that happen."

"There's so much riding on this one man," I said. "It makes me feel a bit better to just talk to someone else about it now and then."

"Sure, we all have to do that and feel like we're double checking things. We know you were the one to turn up all that stuff about the rifle and shells. Thanks for telling us what you know about the cyborg-guy. We think he's OK too."

Our talk did provide me with some of the positive reinforcement I needed. James Stark was from the old school of hard knocks. He had to be over sixty and close to retirement age. There was more than his bald head and the bulge above the belt to tell me that. When a person could see they were going to make it to the end of their career, they usually lightened up some and showed some mercy for those who had a long time to go.

As I was leaving I looked around for the wiry old lady and she was nowhere to be seen, the same as the other times. I thought she would be hard to replace, as she must have had several hundred people and their offices committed to memory. In the future she might be given some treatments to increase the length of her life, or be replaced by a cyborg. I found myself thinking more about such things these days. Where were we headed? If I didn't know that, then how did I know I wanted to go there?

Driving a few blocks south and a few blocks east of the Federal Building put me back to Claire's anomalous apartment. I do remember crossing over a set of railroad tracks on the way, but that's just downtown Austin. It has a number of special idiosyncrasies that set it apart from all other cities. I think it has something to do with keeping Austin strange. We never liked the lock-step sort of thing around here. Claire's allotted parking space was vacant, and I drove on past it to an open space at the rear of the lot. Her spot is close to the entrance. I have never seen any other car parked in it and I follow that lead, unless I know she's out of town. It was still early enough in the morning to be comfortably cool, so I didn't mind the walk.

It went well at my meeting with James Stark and it ended sooner than I had expected. He hadn't given me any cross-check with the CIA, and I wasn't very surprised at that. I had to give it more time. I appreciated his positive reinforcement. As soon as I got inside I flipped on the TV and tuned into a news channel. I could get caught up with the world situation while I waited for Claire.

There was the usual senseless violence in the news, interspersed with the obvious duplicity of government entities. The newscasters liked to have several guests to have discussions, about the same stories that all the channels were beating to death. Some topics they all seemed to steer away from. I didn't think I had ever heard the word Nanotechnology mentioned. When I wanted to do research on the word I had to go to the computer, or pick up a book. It wasn't exactly a household word.

Claire came in at about eleven thirty and we decided to go to a new restaurant that had recently opened up in south Austin. She had an eggplant parmesan dinner, and I selected the chicken fried steak. They both came with veggies and salad. We were so filled up that we passed on desert. We sipped on our red wine and talked a while.

"So, how'd your business work out this morning?" She asked.

"Not bad under the circumstances."

"And, what were the circumstances?"

"Well, I had worked with a government agency and done a few favors for them. I decided it was payback time, but I was asking for too much. The agent acknowledged my accomplishments by giving me something else just as important. He might have taken it badly and given me nothing."

"Ordinarily, I wouldn't pry into it," Claire said. "But I know about all these things you're into on this case like nanotechnology, global warning, developing more water for people in Africa. Scientists say we need to be involved, and I'm interested."

"The people I work for tell me that Bob Clauson's nanotechnology will solve all those world problems. In theory it can do it, but it isn't doing it yet. As any of those things are solved we're all going to know it instantly, and if they're all solved at once we'll know it even sooner with some future shock to boot."

"Where will they do the first molecular manufacturing?" she asked.

"I think it's going to be somewhere around Luling," I said. "That's where Clauson is, and that's why I have to get back there."

"I'm sorry to hear that, Dan. Let's get over to the house for a while, before you have to leave."

We wasted no time getting out to my car, after I settled the tab and left a gratuity. Driving back across Lake Austin to her apartment, I explained how I had to follow the evidence trail back to Crystalville and confront Albert Hoffman. She understood how if it wasn't for that I'd still be in Luling, and when I went back I probably wouldn't return to Austin till the end of the case. I didn't tell her it wasn't just a case, but it was a job I couldn't quit like being in the military. Some things are just too much to tell and I just cooled it. We moved into the ulterior region of the apartment and used the rest of our good time together.

CHAPTER 31

Back on highway 183 I was heading south, feeling greatly renewed and like all was right with the world. I sped along thinking about how I would bring about some kind of resolution to this quandary case. Dealing with big time super-computer companies, government entities, scientists and cyborgs put you into a decoding mode. All of these people talked in their own jargon and raised an eyebrow, so to speak, that was meaning you get my message don't you. Sometimes they even put out slightly mixed messages. I learned to stay cool and just strike a course in the middle somewhere between the mixed signals. I would eventually get things decoded so I could at least understand what was being expected of me.

The decoded message I was getting from CEO, Harold Remkin lately was simply ask Bob to come see us at Gladstone, now and then. It seemed that everybody that wanted to get into the nanotech computer business could do it legally by contacting Remkin. And what Bob Clauson was doing was helping the start-ups get going, but somehow they still were not going. I thought that Comp-ease Computer's was going to be the first one, and then, I thought the fireworks would start with something like a revolution of society in Africa. Suddenly there would be water available everywhere and the New Industrial Revolution would have begun. The world would be put on notice that there no longer needed to be any barriers for mankind. The availability of water, hunger and the other world problems would be solved.

At least the driving was a simple matter, when you could get around mostly on large freeways like 183. It seemed that I had developed routes to travel like a salesman. My travel plans were made easily and I was always relatively close, when you considered what a large state this was. I was getting methodical about it, but I felt better about everything since I had made love with Claire.

Slowing down as I came into the Luling area, I got off the freeway and took the local version of 183 which was Magnolia Avenue. It was the main street but not called Main Street. I hit the traffic lights right and made it all the way to the new industrial park in only a few more minutes. I found a visitor's spot open in the front of the Comp-ease building and parked in it. My last experience in the larger parking lot at the rear of the building, made me shy away from it. As I went through the front entrance door it became apparent that the standard security measures of larger companies were being added here. I was met with a high counter and an intercom, but there wasn't anyone there.

There was a button to push and I pushed it. Following the instructions typed on a laminated card next to the button, I sat down in one of the three black plastic chairs. There was a low rectangular table next to me with recent magazines on it. I selected one of the consumer's report magazines about new cars on the market, and started to read it. I noticed that there was a security camera on the wall behind the counter and I wondered who was watching. In a few minutes the rear door on the other side of the counter opened and Steve Wingate came through it. His speckled-oval-face shined with enthusiasm. He pushed something from behind the counter and the end of the counter where he was standing swung open.

"I happened to be passing by the video screen and saw you sitting there,' Steve said. "I told them I wanted to talk to you and that's how I came to answer your call."

Moving through the open gate, I kept up with him as we entered the regions beyond. Close up I could see he was larger than I had thought, about one ninety, and only a little under six feet tall. There were no robotic appendages I could see on him, so I assumed his sole claim to be a cyber was because of the chip he had implanted in the brain. He motioned me into a side room.

"We can be a little more comfortable here," he said. "We have coffee in the machine, and I have change for us. Will you have one?"

I said I would and he got two cups. We both had it black, and went over to sit down at a table. "I'm sure you're waiting to hear about what came of my trip to Crystalville."

"Captain Torres said you were talking to a suspect about the rifle."

"I got enough information to completely absolve Bob of any blame. He had been held as a captive, by the three men who were killed. Schmidt was paid to free Bob and he did, killing all three of them."

"There's something you need to know about that," Steve interrupted. "Bob developed agoraphobia because of what he went through, they beat him up, and he said one of them was a security guard from Gladstone Electronics. You can see how he was afraid to go back to Gladstone, and started operating on his own. That was when yours truly came into the picture. I think you know all the rest."

"I think I do, now. Has Bob been seeing a therapist for the agoraphobia?"

Steve raised his eyebrows and told me Clauson wouldn't go out of the house. They had to bring the psychiatrist to him and Steve did all the leg work. Ouch!

"Is that why the molecular manufacturing is taking so long?" I asked. "We hear that a lot now," he laughed. He made a contorted face and continued. "Everyone seems to think molecular manufacturing is instantaneous, and it should be, when it's done perfectly with only the purest form of chemicals and materials. What people don't understand, is that even after we know how to do it, there are still more adjustments to make, and we can only get down to the finish through trial and error. It's been tried many times and each time we learned some new refinement that had to be used, no explosions, and there won't be any of them either."

I told Wingate I was reading tech articles from the internet. He assured me that most of the recent information I had been getting by researching internet scientists was correct, but it wasn't exactly a plug in and play type of thing. We mortals do tend to want to over simplify. I should have known that part about how to get it right. You were going to have to do it over and over till it worked.

Finishing my coffee I thanked him for giving me the straight poop. After all, it had to be true. Why would anyone working on such a big thing, want to admit they were reduced to the trial and error method. I thought it was just strange enough that it smacked of the truth. I tossed out the plastic cup and got out of there.

Thinking this would be a good time to talk to Clauson, I drove to Pecan Avenue and Parked. I rapped hard on the door, even kicked it twice. I walked through the gate in the fence and knocked on the backdoor. When I saw that the doors were going unanswered I had an idea.

I got back in my car and drove to the end of the block on Pecan Avenue, into the Pub Restaurant's parking lot. My idea had been right, about none of them wanting to cook. The agent's red Buick was sitting there. They were all inside having a good meal. There was a visual picture I had in my mind of the three agents dragging Clauson out of the house to the car. They could afford to hire a cook to handle meals for Bob. He was supposed to have such a bad case of agoraphobia that he only went to work when it was an absolute necessity. I could see it though, a crisis over the dining room table. Career female agent is determined that she will not be relegated to a home body, taking care of the kitchen for three men. And the men don't want to do it either.

I gave Steve a call back and cleared things with him. Then I called Ray on the hill and told him to tell Willie I was going to start him on four hour shifts for a while.

"I was wondering about you too, Ray. Are you doing OK with eight hours now?"

"I'm up to one hundred per cent now, no problem."

"You know I'm suspicious of the activities that have gone on there. I'd like to keep the surveillance going for a while. You must be about due to make the change."

They were supposed to change the guard in about an hour. I told Ray to call Willie and tell him that his shift wouldn't start for another four hours. It gave me enough time to drive to a spot a little farther away from the place where we had our drop off. I told Ray I wouldn't need a ride. I needed the exercise so I'd park and walk for my shift.

Driving back to the other end of Pecan Avenue I crossed over to Travis Street and took it west to the blacktop road and turned left. I was headed south then, going to the area where we had the drop off spot to go to the hilltop OP, but about a mile before I got to that spot I pulled off and parked on the curb where there were several other cars already parked.

It took me about ten minutes to jog on further south to the bottom of the hill and up to the OP. "Ray had the field glasses trained on the safe-house property. I saw something moving through a window on the first floor," he said.

He handed me the glasses and I took a look, but I didn't see anything. It was a common thing to see things that weren't there, when you were doing surveillance. I had done it many times myself.

I'll keep checking those windows," I said. "You better hope nothing's going on I might have to call you both back here to document things." I told Ray I'd call Willie

"You call and we'll come."

"As a matter of fact, the way this case has slowed, Willie might want to go back to San Antonio. You can talk that over with him. If he does, let me know and I'll ask Joe Torres to assign a couple of his men to help me with the surveillance."

"It's hard to believe how things have changed," Ray said.

Willie's cell rang about five times before he answered it. He said that I didn't wake him, as I had suggested. His car was parked close to his room, and he was checking the tires. I asked him how Helen felt about him being gone so long. He did thumbs up.

CHAPTER 32

This safe-house situation was like a ghost that wouldn't go away. I had started to wonder if it was another one of those mind things. It could be that the image of the place had been so indelibly imbedded in my mind that it kept jumping out at me. See me, I'm still here, better go have another look, it was saying.

Scanning the windows again didn't yield any new discovery, but when I went back to the ground floor window I thought something moved in the outer periphery of the kitchen. If Bob Clauson was there he would know about the surveillance, and might be trying to avoid placing himself in the view of the windows. It was good for him that I provided the security for this place he liked to use now and them. He was aware that if Nelda came here to visit we already had been informed that she had a perfectly legitimate business relationship with him. I had no reason to be trotting down there, the way I had before unless it was for Clauson's protection, not for anything else.

Everything about Clauson's safe-house went together for me now. It was never used as some clandestine place to do business, but only as a private place to meet with Nelda. That clandestine meeting place had been part of my erroneous pre-conceived ideas. I had to strike that out now and clear things up with Willie and Ray. I opened my cell and called Ray. He and Willie were both in their rooms so I got them in a three way conference. I wanted to make it clear about Nelda.

"If either of you are here at the surveillance OP and Nelda shows up to visit Bob you could still call me, because I might be needing to talk to him. But we don't need to run down there and check everything out if it's just Nelda."

"What if Nelda brings someone with her?" Ray asked.

"Don't bother about it if it's someone you know from Gladstone," I said. "If you have any question about it, we need to get into our three way conference and talk it over."

"Clauson knows we're here doesn't he?" Willie asked.

"Yes, and he's aware that he needs security here. He's a hard man to find sometimes. I'm trying to catch him to talk in private, about the meetings with Nelda. Then we can be sure he won't be using this house for anything else. It would greatly help our security arrangement. We seem to be the only ones providing security for him at this safe-house. I think maybe the others don't even know he comes here to meet Nelda. If they do they're prudent about it. In that case I think someone comes with him in the boat or the car and waits outside for him as a sentry."

It was a good thing I had the three way conference, because they both seemed to have the same kind of pre-conceived ideas that I had. We all had a chance to get a new take on it. When we all were on the same page we closed our cells.

I thought about the trial and error method as a basic kind of logic. If you could isolate all the factors that concerned a decision, and then try each one you would make the best decision. When you got down to the last trial you could think of, that would be about a ninety per-cent chance of success. With all the thinking I had done about the factors concerning this house I thought I was at ninety-nine-per-cent. Then I tried to isolate all the factors about Bob Clauson and it wasn't so easy. How did I know that he wasn't living in this house and being dropped off at work, when he needed to.

Going back to my field glasses I carefully scanned the windows. I wanted to be sure someone was there before I made the plunge down the hill, but I could never get the final adjustment on the glasses to see the outer fringes of the kitchen. Then, I thought of the card I got from Steve Wingate. I seemed to remember two cell phone numbers on it and I found it in my wallet and took a look. There was a number for Steve and one for Bob. I tapped in the number for Bob and waited. It rang a few times and then he answered with a cough.

"Excuse me, I have this cold," he said. "This is Bob Clauson, speaking."

"This is Brandon, at your safe-house," I said. "I wanted to be sure about things here. Is this your own personal family property?"

"Yes, it is mine. Are you there now?"

"We're doing surveillance around the clock here. Do you want to keep that up?"

"Keep it up I might use it any time, never know when, so just go ahead the way you have been doing."

"I enjoyed meeting you here that time," I said. "I believe I told you then we had an observation point set up here."

"I think you did. They're always telling me to have more security. Oh, I enjoyed meeting you too. Thanks for calling."

"It's my job to keep up with you, and we can do a better job on the security, if we have an idea of who to look for around here. You might not want to talk about it on the phone. I could come over to your place on Pecan Avenue to talk about it, if you like."

"Yes, that's where I am right now. Can you come over?"

My shift was over in a few minutes and I said that I would be right there. Willie came to relieve me early. I told him about my plan to work more closely with Clauson and make the security arrangement easier to handle. Willie dropped me off at my car before he started his shift.

Driving anywhere, even if you only go a couple blocks to buy something at a convenience store, it takes about fifteen minutes. By using Travis Street as a crossover to Pecan Avenue the traffic lights were timed so you cold hit them all on green. I was at the door of the river-house in about twenty minutes. Bob Clauson met me at the door and ushered me to a private room just off the front parlor.

"You made good time," Clauson said. "One thing I should tell you about that house is about the passage-way-tunnel. We had it sealed up."

"That's the best thing you could have done for the security of that location."

"It was just a historical thing that I liked, never used it much since I was a kid."

"How did you come to use the older one, farther south on the river?"

"Actually, it was a mistake I can't explain. I don't know how my captors knew about it. They had me blindfolded at the first location, so I didn't know where I had been before they moved me to that safe house. I was blindfolded again, but I could tell where I was when they took me inside. They brought me on the boat to attract less attention. I'm not really sure of what happened, when the shooting started but Steve told me about it. I sort of woke up at the river house Steve had rented, after I was rescued from all that abuse."

"Who were the men who escaped through the tunnel?" I asked.

"It was Bud Wilson and Ted Schmidt, the ones that did the shooting that got me free." "I'm sure you know about how that has all played out." I said

"Yes, unfortunately, those two turned out to be criminals who had their own agenda, and you've done an excellent job of police work to prove them guilty of murder."

We discussed all of the questions and details regarding the entire debacle, and asked each other questions and answered questions until we both thought we understood it all. When I asked him who would be coming to his more modernized safe-house, he said only those from Gladstone and himself. He mentioned specifically Harold Remkin, and Nelda Whittington. I wondered why he didn't include Von Mueller, but I didn't say anything. It was his own business who he wanted to have over to his house. I was glad it was a short list. It was almost like saying that it was just a place where he might meet Nelda once in a while. He said he would not be using the older safe-house at all. I asked him if he would call my cell number, to let me know, when someone from Gladstone was going to meet him at the safe-house, and he said that he would.

It was going to give us some breathing room regarding the surveillance. I was thinking of cutting it back to the way it was when Ray and I handled it. That would allow me to go off and do something else while Ray and Willie took care of it.

For the first time since I came on this case, I had some understanding of the principal, and I thought I could trust him. He had been the country bumpkin genius of electronics. Later, I elevated him to the electrical engineer who had made computers small enough to be used at the nanotech level for molecular manufacturing. I progressed on to identify him as a sort of wizard of nanotechnology who might even be able to become invisible. And now I had come to believe he was just an ordinary man with special abilities. He had been the same person all along, but it was my perception that had been changing.

When I decided it was time for me to go, I was surprised to see Clauson chameleon into a person with newly found social graces.

"How about something to drink and maybe a little snack?" he asked. "We have coffee, tea, and soft drinks. The cook's out but we have good sandwiches."

He recommended the ham and cheese, saying he was ready to have one himself, so I joined him. Serving us from the kitchen, himself, he brought in coffee and the sandwiches. He placed them on the claw table that was between our matching claw-footed chairs. The spacious side-room was furnished with antiques that seemed to put me in another time zone, as I relaxed with a breeze coming through the open window.

"You've been through a lot," I said. "I can see how you need the old family home as a special place to go. It doesn't matter if you don't use it often. You know it's still there and you have the anticipation of your next visit."

"Exactly, Steve had a place like it on the outskirts of London he was telling me about, and while I think of it here's an extra key to my house."

"He was a professor of cybernetics there, and he must be an asset to us," I said. "I'd have never known he was a cyborg if he hadn't told me."

"He never talks about it after he tells you, like giving you warning," he laughed.

"You wouldn't want to put him on the spot," I said. "But isn't he sort of using himself as a human guinea pig?"

"He says not, that the chip implant is state of the arts."

"I'm glad to hear it is, but I don't think I'd want one."

"It's supposed to be a permanent implant, never have to dig it out."

I had thought they all had to be taken out after a few years. This was a time with many new materials coming in as a result of the nanotech research. It was an exciting time to be living in. Wingate had said that their next try at the molecular manufacturing would be the one that would work. I was reluctant to mention it to Clauson. It's the kind of thing you cross your fingers over, but I didn't think you should hold your breath.

"Steve seemed very confident about the molecular manufacturing," I said.

"We must be very secretive about it. Our attempts can never be announced to anyone. It's among the top trade secrets at Comp-ease and at Gladstone. I wouldn't think Steve would say much about it."

"Is the security good at Comp-ease?' I asked. "I know the police are dependable."

"Yes, it's actually a kind of cyborg security. Steve has headed it up with some of our own people who are also cyborgs. We also have a fast phone to the police."

"I've talked to John Von Mueller several times about the nanotech research and he answered my questions without giving up any trade secrets and the same with Steve Wingate."

"That's good and after all you're a member of our team, but there have been three occasions, when research regarding molecular manufacturing has failed, because our security was compromised."

"Don't you think the public should be interested, try to get a working knowledge of nanotechnology and how important it is to the future?"

"Oh yes, of course, I'm in favor of that. Steve has been working with several cyborgs, going around to schools to talk about our general plans for nanotechnology. I'm always glad to talk about the generalities, just not the specifics."

"That would thwart security. Just walking in to talk to Steve, I noticed security was better at Comp-ease."

"Harold Remkin is putting in cyborg security at Gladstone too," he said.

Conversing seemed to be one of Clauson's strong points and it surprised me. I thought he was nonchalant about it all, since he had suffered so much because of the failure of security at Gladstone. He was having anxiety attacks and big time agoraphobia, which was the fear of open spaces. I remembered how he was when I came to the door. It had opened a small amount and then further as I came inside. Standing behind the door he was protected from the outside. He must have been listening for the car to drive up, and my steps coming up the walk.

CHAPTER 33

Willie was overjoyed to hear we would know who was going to be at the safe-house, and that the tunnel had been sealed up. It meant no more surprises that would cause us to go scampering down the hill ready to do battle. He wasn't up to all that with his arm in a sling, and I had no right to expect it of him.

It was like a reprieve now that we had won the approval of Bob Clauson, and he was actually cooperating with us. To have that sort of a relationship with an icon was no little accomplishment. I was feeling better about Harold Remkin and Gladstone Electronics too. I owed them an accounting of my time. It was my own shortcomings that had caused me to scoot on by Crystalville and go to Austin.

Making amends became my order of the day, to go to Austin. In only a few minutes I turned back on LH 183 took LH-1 and returned to Crystalville. The polished granite building loomed out over the entire city. Gladstone had been built on the highest point in the city.

The old town was built on the lowest ground, and the new part of town sprang up on the land between the two entities. The old had common names like Main Street, Paint Street and Maple Avenue. The new had names like Bios Street, Atom Street and Silicon Avenue. It also had a wide, impressive road called Drexler Boulevard, which extended from the northern most part of new town to a loop of road that encircled the Gladstone building called Gladstone Drive.

Turning off the freeway onto the access road, I pulled into Gladstone Drive and the parking lot as I usually did. I parked in the first visitors space I came to, went inside and straight to the elevator. It was empty. I pushed the button to the third floor and remembered to think what I always thought when I came here. There was plenty of land around this building to have absorbed the top two floors, and they really didn't need the elevator. I thought that's how the high rise architecture got started and kept going. I got off at three and let myself into the administrative offices. Harold was into his holding the fort mode in the last office. Anybody who came to see him would know everyone else was out, and he was the only one who was indispensable.

"You haven't quit me have you, Brandon?" He said.

"There's so much more than just you in this thing I'm stuck in. No, I can't quit."

"I know what you're yakking at man, things bigger than I am too."

"It's all about Bob Clauson," I said. "And I was only able to sit down with him to talk this very day. He's the principal on this case, and in his pursuit, I've been beat up, shot at and verbally abused. This was the first time he granted me the favor of a small amount of cooperation."

"Hell, you're doing better with him than I am. I'm mailing a check every month to his bank and he doesn't even think about cooperating with me."

"For the first time, I think, I've learned enough about what he's doing to help you."

"Well, for God's sake tell me all you can," Remkin said. "He doesn't even send me a postcard."

"He's got all his table-top-tools set up at Comp-ease to make it the first molecular manufacturing start up. That's the reason he hasn't come back to you at Gladstone."

"Why didn't he set it up here with us?"

"You haven't had good enough security for him here."

"We're putting in the same system they have at Comp-ease right now."

"When you have it in, he'll probably set you up as a start up too," I said. "It's not that he's staying away because he doesn't like you. His friend Nelda asked him to come back to Gladstone, and he wouldn't do it for her either."

"So, it's, because of his engineering special ethics?"

"Yes, he doesn't want to contact you because he knows you'll put it on a personal basis, and he doesn't want to be put on the spot until your security checks out OK."

"Damn it, man, I'm checking up on cyborg security, see what's holding it up," he said. "Go back to Clauson and try to learn more. I'm putting you in for a bonus.

I said that was fine with me and hesitated a moment. He was on the elevator and on his way down, and I thought I could be a spy looking around for a bit of commercial espionage to engage in and turn a few extra bucks. I was all alone up there with all sorts of business records. It was little wonder that Bob wouldn't approve his security system.

As I went down the elevator and outside to the wide open spaces that would have triggered Bob's agoraphobia, I was still thinking about security. What kind of business school had Harold attended that would allow him to be the kind of vacant minded CEO that wouldn't see the obvious evidence of total lack of security? I had walked straight to the elevator, which was still located very close to the front entrance door, and nobody asked me who I was or anything else.

It scared the crap out of me, that with all the great technology this company had, it was all coming to naught. One pompous CEO at the top spot, who had everyone keeping their distance from him, had caused it. There hadn't been anyone else up there with him. I wondered if a new CEO would have to be brought in to fix the security here.

Before I could get back in my car I heard someone call to me from the end of the parking lot closest to the building. Carl Watkins was jogging toward me.

"Hey, Brandon, what's going on?" He asked. "Is Clauson still alive?"

"He sure is, just talked to him before I came here," I said. "I'm thinking what kind of madness is going on here?"

"Remkin madness," he answered. "The stock-holders-board of directors is trying to get him to take his retirement, but he doesn't even have that much sense. I think he's off his rocker."

"I'd think he'd know the company would have to survive if he was going to."

"That's what I mean. He was just saying we needed to speed up the installation of the new security system, and I told him he needed to talk to them more over at Comp-ease. That's the model company for us, but he won't go over there or even send somebody over there. He thinks people are always supposed to come to him."

"Special talent is needed to run that system. Has he hired any cyborgs?" I asked. "They're people who are part human and part machine. They've got a computer chip surgically implanted directly in the nervous system or the brain. They have special skills to communicate with and control the computers used in that security system."

"No, he hasn't done that, because he hasn't kept up with the progress, and he won't trust anyone else to take on the responsibility for him."

"You read and have a computer you use in your office, so couldn't you learn the new system?"

"Yes, I've been trying to learn it."

"The person you should see at Comp-ease is a man who is a cyborg, Steve Wingate," I said. "Would you take on the responsibility for running the new system at least until someone like Wingate can come over here and take it over?"

He said he would do it, if I could get Remkin to assign him to the job. Carl was smart enough to know he would be doing the best thing for himself. I said that I'd go back inside and try talking to Harold. I thought, if I could convince him that he could save things around there and still be in charge, he might go for it.

Tracing Harold's movements through the lady at the console, I found him back in his office, talking on the phone. He put it down and I sat down in the chair next to him, to the side of his desk.

"Listen, Harold I need to settle this with you," I said. "I don't need any special bonus, just for doing my job."

"Well, maybe when the case is over and we figure everything out, we can fix it."

"That will be when Gladstone is set up just like comp-ease to be a start-up company and the new Industrial Revolution will have begun. Our country will be back where it should be, in an honorable position of world leadership."

Harold lit up over the prospect of Gladstone reverting back to their former status as a leading super-computer company. When he started talking about how great it would be to be over this present crisis, I shot my prospect to him.

"You can still pull this deal out of the fire, and get back on track," I said. "I've been over at Comp-ease and have some influential friends over there." The magical words seemed to be influential friends, and I had to bite my lip to keep from laughing at myself.

"Could you get them to help us out?" He said.

I took out the business card Steve had given me and opened up the cell. Like it was an afterthought, I asked. "If they will work with you, and want you to send someone over there, would you send someone or go yourself? Carl told me he had been studying about the new system and would go over there if you asked him to."

Harold said that he hadn't been aware of Carl wanting to learn the system. He thought it was a good idea to have someone ready to delve into it. There was nobody else he had in mind, so it would be Carl.

Punching out Steve's number on my cell Phone, I looked back at Harold till Steve answered. When he did, I engaged in a preliminary bit of friendly conversation.

"It's Dan Brandon, hope your ears weren't burning. I had my evening meal with Bob a while ago and we were talking about Cyborg security, but he didn't give up any trade secrets."

"I like people to talk about me," he said. "It makes me feel important, but I'm glad you didn't get any trade secrets. They wouldn't be secrets then."

"Steve, there's a problem Bob's talked about. I know he was working on getting another start-up going at Gladstone, I'm here now, and they need help getting the new security system set up here."

"We have some equipment over here we could even transfer there, Dan, but we can't seem to get together with them. They will have to have construction changes in the building like we did. They can learn a lot about it by just coming over here."

"Yes, I understand that. Oh, can you hold a few seconds?"

I started up the three-way cell phone conference and Harold started talking to Steve. Then, I made some nonsensical kind of grimace to Harold, like I had heard the call of nature, and ran out of his office. I got to the elevator and stopped to listen to Harold saying he was sending his man, Carl Watson, right over there, so I got on the elevator and went down to the ground floor. As I got out, Carl almost collided with me, and I told him to go up to Harold's office, that he had a new job. His eyes lit up and he wasted no time getting to the elevator.

CHAPTER 34

Gladstone Drive became the loop that took me around to the front of the hotel. I slowed down by the office and stopped at the first unit past it and parked there by the room that was reserved only for special guests.

The key was still on the car seat where I had dropped it, when I thought I would never go back. Nope, hadn't changed the lock on me, so I went on inside and found that there was a tan windbreaker on the chair next to the bed. It was mine alright. I didn't take everything out of there like I thought, and Albert must still be holding the room open for me. He really didn't mean for anyone to get killed and I knew it. Charlie and I had given him a good scare. It probably would not be any more than that. Marsha would never prefer charges against him for the broken window, which had already been repaired.

It was dark outside and I thought about the night when I made Albert drive the car over to his friend's house, and it turned out to be Marsha. I turned the television on and sat down in the plush recliner chair. I kicked back in it and it was good for a pair of legs that felt like they ran a marathon. Something brushed against the door and there was a knock. I thought the sound might be too loud on the TV and I turned it down before I went to the door. "I'm glad you came back to your room, Dan. I know you were disappointed in me." I told him it wasn't something he couldn't live down.

"Just over to Marsha's yesterday, but I'd already laid it all out for her, what I did about that damned rifle. The next day after we were over there telling her about the guys getting shot with it, I went back and told her how I had got involved in it all. Anyway, she's still like family, and she always knew I was kind of squirrelly."

I thought he had been having a few beers, and then he went over to the refrigerator and handed me a longneck. He had stocked my refrigerator too. I took a swig out of the bottle, wow, good and cold. There was a time when I drank a lot of beer like any cop might do on his own time. Shortly he brought me another and continued talking about I know not what. About the time I was kicked back having a third one, he said "I called Marsha, told her you were back and asked her to come over."

"You shouldn't have done that, Albert," I said.

"She likes a beer now and then and you like each other, don't you?"

"Well, yes, but you're not cupid."

"I'm an old man, but I can like to see young people happy."

I heard a car drive up and park close and the car door opened and closed. Her high heeled shoes clicked up to the door. I made it to the door before Albert and let her in. She was wearing a bright colored dress and looked like a movie star.

"I didn't put him up to that, Marsha," I said.

"Oh, what's the difference I came and I'd like one of those longnecks."

I gave her the one I had just opened and pulled a glass out of the cabinet to go with it. She sat down at the small dining room table next to Albert and noticed my pocket watch on the table. She picked it up and examined it, as I pulled another chair up to the table and joined them.

"This is a beautiful watch, Dan," Marsha said. "Where did you get it?

"From a lady friend in Austin," I said. "Don't know where she got it."

"She must have cared a lot about you to give you such a nice gift." She smiled and put it back on the table. I picked it up and put it in my pocket and said nothing. Albert got up and replaced the dead soldiers. He opened a drawer and took out a deck of cards and we played some three handed bridge. We told funny stories while we played and soon were laughing and having a great old time. Marsha won most of her hands and was enjoying the evening. Albert had been worried about her wasting her evenings at home all alone. I could see the parental thing working in him. They were good card players and we played on till early morning. Finally we had all gone to the bathroom, and when Albert came out the last time he just kept going on out the door.

"Albert must have gone to get something," I said.

"He won't be back," she said as she stood up. I got up, thinking she was leaving too. She shoved up against me and we kissed for a long time. "It's alright about the girl-friend in Austin, as long as you're not married," she said. "I like the competition." She pushed on past me and opened the door with an impish smile and closed it in my face.

The next morning I showered and had a continental breakfast, compliments of the Crystal Palace. I wanted to see how it was going at Gladstone, so I walked next door and let the car sit. Carl was swaggering around in the midst of several construction company trucks, with rolled up blueprints under his arms. I could hear circular saws, hammers, and crow bars being used inside. I stepped in carefully, and noticed a new wall was being placed in front of the elevator. You wouldn't be able to come in and go right up to the third floor any more. Other new walls were going in to shunt you to the places where you were supposed to go, when you came in off the street.

Going back outside I caught Carl's eye and he motioned for me to come join him.

"We were lucky enough to get construction people who did the job at Comp-ease."

"How did they get started so fast?" I asked.

"They were still doing some jobs over there and Steve took me to the foremen and they set it up to get some of their other men over here this morning. The foremen came back with me yesterday and we rustled up these building blueprints so they could see the feeder lines and things like that over here at Gladstone."

"That's how they were over here so early with working crews," I said. "I've heard their lingo, time lost is money lost to them. They're in a hurry to finish each job, so they can get to the next one. That's great, working in your favor."

I told Carl how my main concern was still to keep up with things in Luling, since I had met with Harold Remkin and could see which way the wind was blowing. It was a good thing I did report back to Remkin. Things seem to be working out better for all of us now. I could hardly believe that Carl had made a trip to Luling in the late afternoon, and brought back foremen to Gladstone all in the same day. I didn't think it would be hard to put in the special new equipment and learn to use it, once the building was molded to receive the special computerized apparatus used by cyborg security.

They would soon be getting a cyborg or two at Gladstone for key operative positions in the special security system. There were getting to be more cyborgs every day, now that we were really in the space age and not just talking about it. I thought if they needed space aliens to work on special projects they would hire them and run a shuttle back and forth from their planet to ours like gambling casinos pay to have gamblers moved to Las Vegas and back home again.

I was happy about my traveling situation on this case. Most of my driving consisted of only twenty or thirty minute trips. Time was important to everybody. My longest journey was from Crystalville to Luling. It was close to an hour. The largest time loss was changing from LH-1 over to highway 183 and sometimes the traffic on 183 got so heavy that it bogged down in places. I thought if Carl could make such good time I should do even better in the morning, so I bid him adieu and made the trip back to Luling in about an hour. I went on down to the other end of Magnolia Avenue to Comp-ease and parked in a visitor's spot in front. The sun was still high in the sky.

Walking inside, I pushed the button, and followed the short instructions on the laminated card. In a few minutes Steve was there opening the gate for me.

"You get around don't you, Dan?" he said.

"I guess we have to these days." We walked back to the break room.

"Like your coffee black, eh?"

"Yes, you remembered. Thanks for helping Carl."

"He made it back to Crystalville OK, I guess."

"They were knocking out walls and putting new ones up at seven AM today"

"Say, that was good."

"Carl was ecstatic about it, can't wait to get the equipment in."

"Some of it is on the way up there as we speak."

"It's coming from Comp-ease?"

"Yes, in the name of cooperation between enterprises there are freebies, today."

"How can they afford to do that?"

"Special grants for space age science and technology."

"So, has the government been helping the private astronauts to take people on these shuttles to the moon in their own private space ships?"

"Sure they have. I was involved in some of that when I was in London.

Steve must have realized our conversation was going off on a tangent and he switched to a more ordinary vain. He said he was presently involved in a program to encourage more ordinary people to get computer chip implants for the space age and take part in new exploratory space-age experiments. He swore it wasn't dangerous. I thought sure, like it wasn't dangerous in Iraq. We were just going to have the Air Force go strafe around, drop a few bombs and proclaim victory. Well, that was exactly what we did, but years later we had still been there, and it was dangerous.

"Are there statistics available about the history of chip implant safety?" I asked.

"Sure what would you like to know?"

"The outcome of a ten year study of how people turn out after the implants."

"I haven't heard of anybody having any problems after the surgery."

"Oh, see now you're calling it surgery," I said.

He bellowed out laughing. "You're starting to remind me of someone," he said.

"Who is that?"

He laughed another short jag, and said, "Woody Allen."

"I can't object to that," I said. "Woody is one of my favorites."

"I didn't mean you should become a cyborg," he said. "You're doing more good just as you are. Do you think I couldn't tell who was really behind getting Gladstone back in the picture?"

"And I didn't mean anything against the cyborg promotion, either, it's just not for me. I know we do need more cyborgs and I'll do all I can to promote it.

"Right, I knew you would, so I told you about it, and it's alright to feel the way you do. Many others feel the same way, and we only want the right ones. It's not for everybody.

"Since you mentioned Gladstone," I said. "Bob was talking about how you both owned some family property and were proud of its historical significance. He's going to continue to use his safe-house, but he said he'd like to use it only for special meetings with his friends from Gladstone." Steve smiled at that.

"We'd rather have him entertain them at the plant, where we have the new security, but I've been going along with that for him. I still appreciate good old fashioned surveillance as a good form of security, and I know you and your operatives have been doing an excellent job of that."

"Bob said he would let me know when he would be there from now on." I added my thanks and told him I thought it would be a big help.

CHAPTER 35

Steve got a call on the intercom phone in the break room and took it there. He said it was about some other equipment they were ordering for the new system at Gladstone, and suggested that I could go back the way I came and let myself out. It was his way of showing me, that as far as he was concerned I was just another trusted member of his team. I appreciated the confidence, let myself through the gate with no problem, and walked back outside to the car.

Driving the short distance to the southern outskirts of Luling and the room I still had at the motel, I pulled in at the far side of the parking lot, where there was more privacy and parked. I wanted to check with Charlie Sloan, since I hadn't got around to it when I was in Crystalville. Wanting to satisfy myself that Albert was really off the hook with the CPD, I punched in the number for the detective's room and in a few moments Chris Amani answered. I asked her if she had heard anything from Charlie about Albert Hoffman and his involvement in, what we called the deer hunter killings. She didn't think they were still following that, since the murders had been solved. She said Charlie was there and I heard her ask him about it, then he came on the phone.

"Haven't seen Albert lately," Charlie said. "I can't say I like him for anything, damned poor criminal." He could hear me laughing and started to laugh it up himself."

"I tend to agree with you, Charlie," I said. "Albert's not likely to get dirty, with his lawyer following him around." We had another good laugh and I heard him say something to Chris.

"Oh, Chris has been missing her friend Ray," he said.

Charlie wasn't pressing anything about Albert. We joked a few more moments and then we closed our cells.

I closed and locked the car and carried a few things to the room. I put the wind breaker on a hanger, and the other things I dropped on the top of the chest-of-drawers. I could hear Willie moving around in his room next to mine, so I went over and filled him in regarding security at Gladstone and Comp-ease. I told him we didn't need to have the safe-house under surveillance at night, since Clauson wasn't going to be using it, except for an occasional meeting with someone from Gladstone. I had also discovered that during the night time police cruisers drove by river properties every hour. We didn't need the duplication of services.

"Wingate is sharp about security," Willie said. "I noticed that when Comp-ease first opened up and we had them under surveillance. He walked over to my car two blocks away from their building, and introduced himself. Didn't tell me he was a cyborg though."

"He's an expert in robotics and they're using a lot of that in the new security"

"I read, listen to the news on TV and try to keep up with technology, but it's hard."

"There's a lot of cultural lag, today," I said. "New technology is coming in faster than people can learn it. We just have to help each other and not be afraid to ask questions."

"I've got one right now. What do they have to do to be a cyborg?"

"They've had some sort of surgical transplant of organs or other improvements, such as a pace maker or a computer chip implant."

"Why do they want the computer chip implant?"

"It allows them to interact with computers and robots more efficiently."

"So, they're part computer themselves, with the chip."

"Yes, some people get the implant to be more employable," I said. "Wingate is starting up a new program to educate people about cyborgs and promote it."

"You mean they're doing that at Comp-ease?"

Telling Willie we'd be helping ourselves to learn more about cyborgs, I said I'd relieve Ray on his shift of surveillance and they could go over to get information from Wingate that we might hand out to interested people. I drove over to the place at the curb where I had parked before, parked and walked over to the surveillance spot. I stopped at the tent to get a bottle of water and hiked up to the top of the hill.

When I filled Ray in about it all, he was also interested in the cyborgs. I told him that Willie would be waiting for him at the motel room, and they could go over to see Steve Wingate together. He liked the idea and went down the hill to put it into action.

Clasping the binoculars, I adjusted the lenses to the right focal point. Training them on the windows, I renewed my interest in the historic property. I remembered how Ray had imagined seeing someone moving around in one of the ground floor rooms. And then not long after that I about had myself convinced that I had seen something myself. Instead of focusing my eyes too long and falling into that trap again, I sat down the glasses for a while and just enjoyed the scenic view. It was a bright clear day and you could see for a long way. There were a lot of birds, circling around up there. I thought one of them was so high that it must have been an eagle.

One of the cars I had seen before drove in the driveway of the house where I had thought government agents were doing surveillance. A woman I had never seen before got out of the car and went inside the house. Ray had thought he saw Tom Cobb drive up to that house and go inside. I wondered how he could be sure about it from that distance, since he had met Cobb only once. I had accepted it as the FBI house, but now I wasn't so sure about it.

I picked up the field glasses again and went back to scanning the ground floor windows. I thought I could see more clearly in the kitchen than I had seen the other day. On such a bright day like this one, there wouldn't be as many shadows to assist your imagination.

It didn't seem like it could have been even four hours and Ray was back to relieve me. They were just doing six hour shifts now and that was enough. I had as much of an imagination as any one else. When you were watching close like we were doing it wouldn't take long for your eyes to be playing tricks on you.

"You see anything down there?" Ray smiled.

"No, not a thing," I said. "Learn anything about cyborgs?"

"Yes, we learned a lot, and we got handouts in Willie's room."

"Did Steve give you any demonstration of his special ability?"

"He took us back to the production area and all the doors opened for him."

"Wouldn't they open for you too?"

"We tried it but they wouldn't open for us."

"Was there anything else?"

"There were two cyborgs on the assembly floor, a man and a woman. They did the job by telepathy. There was a special computer that was supposed to have a million answers and formulas to give you if you could ask the right questions. We thought of some questions and wrote them down. The cyborgs just looked at the question we had written, sent it to the computer by telepathy and got back the right answer. The cyborgs could send messages by telepathy to each other too."

Ray also mentioned that there were many cyborgs involved in space exploration, and there was even a special space colony for them. He said Willie would have literature about all of these things back at his motel room. I planned to check it all out later. It was quite apparent that I couldn't learn about all of the technical innovations of the day, even if I could devote the rest of my life to it.

Leaving Ray in charge at the OP, I descended from the wooded area to the black-top road and found my car at the curb where I left it.

I drove back to the motel, parked, and went to Willie's room. I looked over the hand out material for a few minutes, selected an article about cyborgs in space, and took it to read in my room. There was only one chair in his room and he was using it to read one of the hand outs himself. We exchanged a few witty remarks. I went next door to my room to get comfortable in my own chair, read a while and quit being witty.

Well they say one picture . . . It showed the cyborg working on some kind of space dock with the space ship in the background. The cyborg was doing something with a robotic arm attached to the space dock and it occurred to me what a great asset these people were to the space program. They could do a much better job, being so sensitive to computers and robots and be able to send and receive messages through telepathy. The computer chip implant should be in the central nervous system, close to the brain or preferably in the brain for there to be good telepathy. I assumed being telepathic would be a great asset for an astronaut, and I thought before long they would all go for the computer chip brain implant. It was a highly competitive program. I could see how it could be added to the long list of requirements for those people. It was a good thing that they were paid well.

Willie thumped twice on the wall and I thumped back, our signal for the visiting mode. He came over bringing the light kitchenette chair, and then ducked outside again for a few moments. He returned with two plastic cups full of coffee. I flipped on the TV and turned down the volume so we could still talk over it.

"I heard earlier on the news, about the moon shuttle business," Willie said. "No accidents for any of the companies doing it for the last five years."

"You, want to bet they employ only cyborgs?" I asked.

"No, I won't take that bet, I'm starting to think everybody is going to end up with that brain implant."

"They may have to if they want to get the better jobs."

"Some law enforcement groups want to implant an ID chip in every citizen so they can track criminals, especially the ones who committed sex crimes."

"I've heard they're already doing it on some of the worst ones," I said. "The ones that move around the country and refuse to register themselves when they arrive at a new place really deserve it."

Willie was a hard line, old timer with most law enforcement issues and he didn't waver with this one. He was highly in favor of the ID chip for everybody. I would go along with it only if we had sufficient evidence that it would help strengthen the American family and protect children. There were still far too many cases of child abduction, and murder. One case would be too many. Legislation had been enacted that was helping. We still needed to do more.

I thought there could be more cyborgs with the power of telepathy employed in key positions in law enforcement. If they could read the minds of the sex offenders and track them, some of the crimes could be detected and the criminal apprehended before the crime was even committed. That would not only be efficient, but also cost affective. And as for general police work, can you imagine the effect upon a criminal, when a cyborg reads his or her mind, and brings out all the details of how the crime was committed and by whom. Immutable facts delivered directly from the criminal's mind would be soon converted into the kind of evidence that would either convict or acquit the suspect.

CHAPTER 36

It wasn't that I was waiting for my cell phone to sound off, but thinking about telepathy made you more aware of such things. For some reason, I had been toying with my cell phone and I checked it to be sure it was on. When it rang I had to think that I had the power. I had been thinking about the phone. Yeah! I knew the sucker was going to ring and it did. The caller was Steve Wingate.

"We were to let you know about anything going on at the safe-house," he said. "I'll be going by there to drop off some groceries. I check the place out now and then to see if anything is needed there. I'm going to be alone, and I'll park in front."

"Does Bob plan to have a visit?" I asked.

"Oh, no, just a standard maintenance thing. If he does we'll let you know about it."

"Yes, Ray is there. I'll call him now. Thanks for calling." We closed our, cells and I punched out Ray's number. It rang about six times before he answered it. "I'm glad you're, OK, Ray" I laughed.

"I was checking out a sound in the brush, some little vermin," he said. "That's why I didn't answer right away."

"Wingate is going to be dropping some supplies off at the safe-house. He'll be alone and will park in front, so you won't have to worry about anyone being at that place on the left side or in back. When he leaves you could close up there for the day, and come join us at my room. We're having a nanotech and cyborg meeting here. I'm serious about it. We've been reading those hand-outs and talking it up big time."

Ray said he'd pick up coffee on the way and we closed our cells.

"What was the problem Ray had with the cell?" Willie asked.

"He went to investigate a noise in the brush and left the cell laying on the tarp."

"I'll tell him to put it in his pocket next time, when he gets here."

Willie had brought all the other hand-outs with him and we continued to read through them and occasionally talk about them. When Ray came into the room we continued to discuss the high tech topics, until we had enough for a while and disbanded our meeting.

An image had been trying to break through in my mind, and I was doing what I did at such times. I sat back in my chair and imagined a blank screen. It was a technique I had read about, when I was a young man considering a career as a magician. I don't know how scientific it was, but sometimes it worked. I had believed in ESP long before I ever heard of a cyborg. I was young and impressionable in those days. I attended magician shows and read everything I could find about the great magicians like Houdini.

One of the things I had tried to do was to walk by a department store window, take a quick look and write down all the things I could remember about it, after I had gone past it. The next time I went by that store I'd take out my list and see how well I'd done. It seemed that the more times I did that mental exercise the better I got at it. With different store windows used each time, of course. I never cheated. I still do it sometimes driving around an area of town where new buildings are going up, in an area that's important to my case. I try to remember what they look like and the type of business they would be used for.

Each time I went to the Industrial Park to visit Comp-ease I had noticed the new buildings. The last time I had been in a bit of a hurry and as I was driving out I caught a glimpse of a building where the construction had been completed, and equipment was being unloaded from trucks and taken into the building. I thought that the image I had been trying to remember was probably the sign on the front of that building. My ESP wasn't working, so I drove back over to the Industrial Park for a closer look.

The sign in the front of the new business read Gemini Computers, and there were a few words added to the company logo that claimed it was the space age computer. I thought that would mean they were in the smaller, super-computer category, like Comp-ease and Gladstone. It made me even more curious as I paced around the front and pondered over the high-tech atmosphere. I had driven completely around the modern red-brick three story building. It was larger than the Comp-ease building. The equipment going in the back-dock had bulk and weight, so it made sense to me that the product would be small. For small accurate measurements to manufacture something small you needed a level solid-concrete floor to hold the large equipment that is bolted down to the floor. That was something I learned, when I was a skinny kid working in a machine shop on a part time job.

Curving around a manicured turf in front, there were marked off parking spaces for some of the executives and visitors. Three late-modeled cars were parked in executive spaces, but most of the spaces were for visitors. I had parked my used flivver in one of those spots, and was still standing on the sidewalk by the front door, when the bright-eyed oriental man got out of his car with his briefcase in hand and started walking toward me.

"Are you with the news media?" He asked.

"No, but I'm interested in the new business going in here."

"Oh, are you in the electronics business?"

This intelligent, soft-speaking executive was doing his job, checking me out in a pleasant, respectful manner. It was for me to respond in kind. I handed him my card.

"I'm working for Gladstone Electronics doing some security work for them."

"Yes, I understand," he said. "We are putting in a good security system, like Gladstone Electronics has. I'm John Chin CEO for Gemini Computers.

"I'm Dan Brandon. Would you know, Steve Wingate?" I asked as we shook.

"Yes, he's helping us to put in our security system," he said. "I will keep your card for security. Sometime we might work together.

"You made a little joke," I said. "It'll make you feel secure to keep my card." He nodded and we both were laughing as he continued on inside and I got back in my car. I turned the key in the ignition, drove out the back way and farther south to my motel room on the edge of town.

Thumping on the wall brought Ray and Willie back for a short meeting to pass on what I had learned about the new computer company. They seemed impressed at how I'd been keeping such a close watch on the growth of the small industrial complex.

"Sounds like it could be the third nanotech start up," Ray said. "Since Wingate is working with them to set up the security."

Deciding to clear that up, I called Steve on the cell. He answered on the second ring. I told him about my chance meeting with John Chin and what I had learned. It occurred to me that he might resent my nosing around, and I wanted him to hear it from me, before he heard it from John Chin.

"Yes, we are helping Gemini to get started," he said. "But as slow as Gladstone has been, I don't know when they'll be ready to start up. John Chin might overtake them, and Gemini might be the second one. They'll go to second and Gladstone could fall behind, to third."

It sounded like he was cracking wise, but he wasn't laughing and I was sure he was just being matter of fact. He had always had that kind of an attitude and I liked it.

"Our motel is only a few minutes away from the industrial complex and I've been watching the new buildings as they were going up. I hope you don't think I' m too much of a meddler."

"No, I don't think that," Steve said. "I'd always rather have people be interested in things and keep a close relationship."

"That was how I thought you felt, but I just wanted to be sure."

"There's no problem with folks like you hanging around as much as you want. It's more security for us, when you're aware of what's going on around there, so keep it up. What we worry about is the wrong people being there, for the wrong reasons."

"If I run into any of that kind I'll let you know and call the FEDS."

"I'm sure you would, and we appreciate it."

I thought the wrong people being anywhere for the wrong reasons would always be trouble. It was good to be clear about all this with clients. It was a two way street and both parties had to be considerate of the other's situation. It changed too, and you still had to keep up with the changing situations. On that note we closed our cell phones, and I got back to the discussion with my operatives.

"So, he didn't think you were meddling around there," Willie said.

"No, he likes us to be that way. The more we know helps their security, but it doesn't hinder it. They won't tell us trade secrets and we don't want to know them anyway."

"Did he verify that the new company is another molecular manufacturing start-up company?" Ray asked.

"Yes, he even said that he thought Gemini was going to be ready before Gladstone."

"So there are three start-ups that we know about now," Willie said.

"Yes, and Gemini advertises as the space-age computer," I said. "That might imply that they manufacture computers to be used in space travel."

"I can see how NASA standards for computers would be high, when they're going to be used on spacecraft," Ray said. "There would be requirements for them to be able to stand up to special vibrations and temperatures."

"The technology keeps changing too," Willie said. "Before they could have people shuttled back and forth to Mars they had to solve the radiation problem. They started out by shielding astronauts from radiation by hitching rides inside of asteroids, and added other special techniques that solved the problem."

"It's good for the economy," I said. "As they travel to planets they have to employ more people to make the special equipment, like we've seen with Gemini."

I didn't mention how they had to employ people like us for special security problems. It had been stimulating the economy for some time now and we were up to full employment. Only one per-cent of the federal budget was used by NASA in the space quest, and the rewards we've reaped made it a great investment. Special new materials developed through nanotechnology haven't been just a help, they've been a necessity. New patents and new companies have come out of the research to also help the economy. And one of the outcomes of the nanotech development of molecular manufacturing would be to have spacecraft available to initialize the colonization of other planets. We would be able to duplicate and mass produce every item on earth after the spacecraft had landed on the chosen planet. Each spacecraft would be like a Noah's Ark to replicate our culture and our people on other planets.

If there were other living beings similar to us on them, they might look into the heavens and see us coming. We would be the aliens from space. I thought we had been concerned so much about aliens coming to earth that we should have empathy for the inhabitants of our new colony.

CHAPTER 37

"If we had been launched from a spacecraft," I said. "And we were landing on another planet how'd you feel about finding other people on the planet?"

"Wow, I'd be frustrated," Ray said. "The inhabitants could be suspicious, and might even attack us."

"Hope I'd be going to Mars, since they're already started there," Willie said. "That colony is underground and they explored all around the underground water supply without finding anybody else."

"Mars would be good from that standpoint," I said. "But what about an asteroid headed toward you? On earth we had help from the U. N. since 2009, to prepare a Gravity Tractor. We hope the Apophis asteroid will deflect it away from us by April of 2036."

"Our little colony on Mars isn't ready to provide all the things we have on earth," he said. "I guess I'd just hope we were far enough underground, and the asteroid wasn't one of those big ones that could take out an entire city."

"All of those things have to be thought out, and plans made a long time in advance like NASA does," Ray said.

Our discussion brought us to face the reality of space travel and space colonization. The colonization of other planets was just one of the basic concepts of nanotechnology. Interplanetary travel and colonization was a necessary part of the over-all plan. It was a bit of a surprise to many of us, as the advanced technology unraveled, to see how much more was yet to come. We couldn't go back to earlier times we might have liked better. It was not an option. Time kept marching and we had to go marching along to the same tune.

Returning to our present dilemma and what we should do about the specter of anti-American sentiment, being directed against NASA, and our leadership in the fields of nanotechnology, I focused our attention on those things.

"I've heard a lot of complaining about money spent on space exploration," I said. "There is only one per-cent of the annual budget spent on NASA, but there are always those who have there own ideas of how the money would be better spent somewhere else. What do we say to those people?"

"I'd say, since we haven't solved the earth's main problems," Willie said. "We needed a back up plan. We haven't solved global warming, energy conservation, or the endangered species, so we need to be prepared to take our people somewhere else if we had to. Nanotechnology is the back up plan. It could provide us with a modern, kind-of high-tech Noah's arc."

"Well, I wouldn't know what to say," Ray said. "Nanotechnology is supposed to be the back-up plan, but why is it taking so long to complete it?"

"You mean to accomplish the complete molecular manufacturing?" I asked.

"Yes, I've heard the top scientists in the field of nanotechnology have different opinions about it. Some think that the ultimate copy of molecules won't work, as you would make a copy of a document on a scanner. Some of them don't agree with the table top molecular manufacturing. They have to agree it can be done with many types of materials, because that has been done for a long time. But they don't think we will be able to reproduce just any object and then mass produce it."

"They don't have much argument, though, with the idea that it will eventually be achieved," Willie said. "I think that's where they've got us. They keep saying we're right at the edge of the technology and ready to do it any day now."

Giving a brief summary, I pointed out that since we weren't the scientists, and engineers doing the research nanotechnology we had to trust them to do the job. Most people felt the reward would be so great that it was worth waiting for, and these scientists and engineers were already on the payroll at some university or large company anyway. What's the harm of a few scientists and engineers tinkering around with it for the rest of their life, and maybe their children, and their children's children tinkering too. After all the world was wearing out, and we couldn't afford to pass up any way to save it. If it was going to be saved this would be the main group we had to depend on to do it.

"Our nanotech people aren't just tinkering around," Ray said. "They seem enthusiastic about it, and that makes me think they're going to do it."

"There's something a bit off about this case, though," Willie said. "We were hired to find Bob Clauson, and when he turned up Gladstone still wanted us to spy on him. The pretext was that we were to try and get him to return to Gladstone. And now we seem to be just extra security. How will we know when this case is over? I can't stay on it for a couple of generations while they perfect molecular manufacturing."

Telling my operatives from San Antonio that I had kept a list of the activities they had been engaged in including the medical bills, I assured them that I would type it up and turn it in to Harold Remkin at once so they could get paid. They were both ready to be cashiered out of this questionable outfit and I didn't blame them. I called Remkin and he said he could have their checks ready today if we wanted to come up there to get them. I said we would and if there was a bonus to divide it between them. It paid to test your boss once in a while and in my estimation Harold Remkin had passed the test.

When I told Ray and Willie that Remkin would have their checks ready today, they went to their rooms and started packing, so after I formally typed up their report with bills and receipts they'd be ready to take my signed reports with them to Crystalville. I had kept all of their individual bills and receipts in two separate plastic, baskets. I typed their individual expenses and accounting of things accomplished and I was liberal when estimates were required. It did surprise me that my two operatives were so quick to go for the entire one hundred yards and be cashiered out.

They came back in the room in about fifteen minutes, with their luggage packed and ready to go. I had my typewriter set up on a card table, and was just finishing the typing. I signed both reports and handed them over for them to give to Remkin. "Keep in touch." I said. Willie looked at his report and smiled. "Did I do all that stuff?" he laughed.

"You did it all and more." I said.

Willie was always modest. Ray didn't even glance at his report. We shook hands and they were out the door. I looked over the items that were in my own basket along with the pages of my daily account of my activities. It would only take me five or ten minutes to type it up also to give to Remkin. I considered doing that and getting into my own car to take my own report to Crystalville and be cashiered out along with my operatives. It didn't seem to be the right thing for me to do, so I hung in there and hoped for a break. I thought if I made a few inquiries it might clarify my thinking, and help the decision making process.

I called James Lebeau, the executive officer at Comp-ease, and he came on for a few seconds and asked if I could hold. I said yes, pulling out my wallet to fumble through it looking for other business cards I might need. He was back on his cell in a few minutes, apologizing for the delay.

"I just wanted to touch down with you," I said. "I've been working closely with Bob Clauson and Steve Wingate lately."

"Yes, I've been hearing you've been quite a help and I want to thank you."

"Oh, I've just been doing my job, but that's what I called about. My operatives had to go back to their regular jobs and I have questions about my own role in this case."

"I appreciate that, and I'd like you to know if you have any problems with Gladstone I'd be glad to hire you to do the same things for my company."

"Thank you, I'll keep that in mind, but it's not about problems with Gladstone, it's more about the success or lack of success of, molecular manufacturing and how I can know when I've reached the end of this case."

"Now, I understand your concern," he said. "We need your help on a lot more than for the security of Bob Clauson and the start ups."

This came as a shock to me, and I asked what other things he needed help on. That was when I got the big shocker. He told me that he had received information regarding another computer manufacturing company. This company was purported to be at the nanotechnology start-up stage, but not under the jurisdiction of Bob Clauson. He wanted me to investigate it, and gave me the address and name of the company. I wrote it down and said I would go there and check it out. We closed our cells and I pulled out my map. It was on highway 80 at Martindale another company out in the boonies.

Taking 80 north out of Luling I drove about fifteen miles to the small industrial area. The name was Vansar Electronics and it was the only computer company located in that area. Two other buildings were in the small complex, a paint company and a chemical company for swimming pools.

The Vansar Electronics building was a modest one-story-wood-frame structure with thin composition shingles that looked like they'd blow away with the first little twister. The slope angle of the roof appeared to be only a few degrees, and it almost qualified as a flat roof, like I'd seen on some storage buildings. There was only dirt on all four sides of it. I parked near the door and went inside. I was met by a middle aged man in tan shorts and a tee shirt, who said he was Ed Tanner.

"Are you building computers here?" I asked.

"No sir. They're fully assembled," he said. The way he kept glancing out the window and back, reminded me of men who had recently been released from jail.

A display case contained computer specifications and the assembled Vansar Computer. I looked over some of the specs and noticed it was gigabyte technology, not nanotechnology. I could not imagine any molecular manufacturing being done here.

"Could I have one of your business cards?" I asked.

"I'm having some made up, don't have any now though."

Reading the literature I had taken, I saw where it said that the computer could be custom made. I thought that was unusual and questioned the man.

"It says here," I said. "These computers can be made to order. I'd like to know, where I can go to have one made to my specifications."

"I don't know where they're made, but if you give me your name and phone number I can have somebody call you and make an appointment with you.

"Well, you must know someone who comes around to keep you stocked. Can't you give me a name?"

"No sir, I can't. When I get stocked it's a delivery service."

It was getting uncomfortable for both of us, and I was convinced that I wasn't going to learn any more here. I gave him my right phone number but a phony name, so I might still have a chance to catch the one who called me. He said somebody would be calling me and I went back outside to my car. The wind was strong enough to blow dust in my face as I got in the car. I was glad to drive away from this dirt bowl. It seemed like a front for something more sinister.

CHAPTER 38

It was late afternoon, when I drove back to Luling on highway 80 and found James Lebeau still in his office. Wanting to make personal contact with him, in the same day he told me about this new company, I hoped he might drop more facts to help me in my investigation of Vansar Electronics.

"I wondered where you heard about Vansar," I said. "It's in a real mud hole and the proprietor wouldn't or couldn't even give me a contact name."

"I found out about if from a friend I went to College with. We ran into each other at the conference in Austin. He's an electrical engineer with Aster Computers in Lockhart. We exchanged business cards, since we're both in the computer field trying to keep up with technology. We've talked on our cells about every day since then. He called me yesterday to warn me that some electrical engineer with a doctorate was making nanotechnology claims. The man was trying to sell his ability to set up a table top molecular manufacturing unit. My friend didn't know that we've already gone into it here at Comp-ease."

"So, what is the connection between Vansar and the man trying to sell his nanotech skills?" I asked.

"He said they have it set up there at Vansar and have already mass produced objects with a more high powered type of Vansar. If you can take about any object and mass produce it with molecular manufacturing it will cost a lot less than the old manufacturing process."

"Did your friend get the name of this nanotech wizard?"

"No. It's a little more than a rumor though, it seems, since you have already verified that something's fishy about Vansar. I suspect they have some salesmen going around talking to the super computer companies who are into the very small computers."

Saying I might need to contact his friend, I asked for the name and phone number and John gave it to me. I wrote it down in my pocket note book. I checked my cell phone to be sure it was on in case I got a call from the Vansar representative. It wasn't clear to me, if there was something criminal going on at Vansar. It seemed that James Lebeau thought there was a conspiracy to defraud, but he couldn't point to anything as proof of it. I told him I planned to respond to the phone call, by placing an order for a special nanobyte computer. The order wouldn't be taken over the phone. I'd be told to go someplace, where a representative would talk about money, and be ready to take a check from me.

It seemed that Vansar was a venture started on a shoe string. Standard marketing procedure wasn't being used to sell these computers. I thought the location I had visited was only a dummy store front, and the man was more like somebody just putting in his time, than a salesman. If they didn't want to sell them, then why bother producing them and putting them in any kind of store. That was when it came to me. If you reasoned by trial and error, as Lebeau had undoubtedly done, you would see the conspiracy to defraud as the only reasonable answer. Why else would something as expensive as a computer be produced? The rules of marketing were being ignored, because this wasn't the real product they were selling. I would have bet there were only a few of these computers turned out in someone's garage or basement. The over-educated, nanotech wizard had found a guy on the street, and had given him the lines to say. If I had one of those computers that could be the proof we needed.

"The computer is how we can close that operation down," I said. "Would you be willing to pay for one of them?"

"Yes, I would, if we could prove it was a fraud," Lebeau said."

"The Luling police have an up to date fraud division. I'll talk to Captain Torres and get one of his plainclothes detectives to go back there with me. I think we could shake up the man tending the store."

Lebeau gave me one of his company's blank checks and I drove three blocks to the LPD and parked at the curb. I went in the front of the building and found Joe Torres in his office. I told him about the suspicious set up at the nearby store and the story Lebeau's friend had given us. Since I had already ensnared Joe into the nanotechnology business, he wouldn't assign a man to go with me. He pulled himself up straight to his over six feet in height and went with me himself.

Joe took an LPD van with a caged in rear area for prisoners. I thought it would make a strong impression on ex prisoners. He didn't put on the siren as he didn't want to make the man bolt out of the store. I followed him in my car. Joe parked close enough to the front of the building, so you could look through the window and see the LPD van. I parked behind it and we went inside. He was going to let me start things, then chime in.

Taking the literature I'd picked up earlier out of my pocket, I found the laptop model described as smaller design, grater information storage space. I handed the literature to the man and pointed it out to him. He took a look and handed it back to me.

"I'd like to buy that one," I said.

"That's only a special deal," he said. "I don't stock it."

"You mean you won't sell it to me?"

"I didn't say that, you've already special ordered it, when you were here before."

"You said somebody would call me and nobody did," I said. "How could it be ordered when nobody called me back to take the order?"

"I put it in for you, since you'd written it down with your name and phone number."

"You told me you didn't have a phone number for me to call, but I see you did."

"The company doesn't let me give out that number," he said. "I'll just be a few minutes, sir." He said to Torres.

"I'm with him," Torres said.

"Oh, fine," he said as he glanced out the window and saw the LPD van.

"Well, can you give me the price of that laptop?" I asked.

"He told me it was over five thousand dollars but it's only sold to corporations along with some other stuff designed for the special order deal."

"It's the he, we want to talk to," Torres said.

"Yeah, that's your van in front."

"I'm Captain Joe Torres, Luling PD," He said, flashing his badge. "You can cooperate with us now and go on with your business, or go with me over to the station till you're ready to cooperate."

"Yes, sir, I'm ready," he said. "Here's the phone number." He reached behind the display counter and gave Joe the piece of paper with the number on it. I wrote it down in my pocket notebook.

"Now, you can tell us the rest of it," Joe said. "Start from the beginning."

The man was trembling and pathetic. The one who had set him up in the business was a big, well-dressed man with brown hair, who had talked about football and seemed like he had been a player. We tried to get the store keeper to remember facial features, but all he could say was that the man who employed him had an average, well proportioned face.

The store keeper said his employer had hired him to replace a teen-ager the day before, but he had never seen the girl. The complaint against her was that she had been on the phone with her boy friend every time her employer had called to see how things were going. The boss said he couldn't spend the time calling half the day, till the line was not busy. He paid the new man a week's pay in advance, and it proved to be a mistake because the new man was more willing to cooperate with us. He had enough money to see him through to his next job. As soon as we were convinced that he had told us everything he knew, we would leave. First, Joe wanted to be sure we hadn't missed anything.

"What was the girl's name," he asked.

"He didn't tell me her name or anything else, except he was going to try older people now, no more teen-agers."

"So, did he tell you to keep the line open?" I thought to ask.

"Yes, that was when he paid me in advance, and said he'd not be coming back if I did well. He would know how well I was doing by calling me. I would get my next check by mail and if I did what I was supposed to do, the next one would be a week in advance too."

"Well, sorry to ruin your sweet deal," Joe said. "We have to find him any way we can, and I think we have some funds to pay you for the next week. How much he pay you?"

"He gave me three hundred in cash."

"No tremendous amount, so we can match that, with a small bonus at the end. This is when you show me your driver's license."

Our man fumbled with his wallet. Some of those guys came out of the joint with no valid driver's license and never got around to getting a new one. I was relieved when he pulled it out and handed it to Joe. That was one requirement that couldn't be waved if he was going to work for us. Joe looked at it carefully and wrote down his name. I was hoping Joe would assign a man to watch this place for a while and give the name to him. I wanted to be free to try some other things to locate the nanotech wizard.

"Tell you one more thing before we leave," Joe said. "When your boss calls you be careful you don't give yourself away. Just play along with him. I'm sure he's running this scam on his own." Joe handed the driver's ID back to him and stepped away from him toward the door. He stepped outside and made a call on his cell.

The man looked away from me and made himself busy dusting the top of his sales counter. He knew I was some kind of cop and I left him alone, while I waited for Joe to finish. It took longer than I had thought. He made several phone calls. I was watching him through the window in the door, and listening for the phone inside to ring. Joe put his cell phone away and motioned for me to come outside where we could talk privately.

"I had a trace done on that phone number and sure enough, it's a residential phone in Martindale. It's only a few miles from here if you want to check it out later. For all we know the guy might be using his own phone."

I said yes, I would check it out. And Joe said it was in his jurisdiction and told me how to get there.

"There's a plain clothes officer coming over here to watch this place, and we'll be talking to Jim, in there a while, to get him started out right with us."

"I really like the way you do things," I said.

"I thought you would." He smiled and went back inside, and I got in my car and drove into Martindale.

CHAPTER 39

Climbing out of my car to look around, I noticed a housing development had sprung up in the land past Main Street in Martindale. Twenty years ago Main Street had been it. There had been nothing but green past it. These houses were not crowded together into a small division and you could see green left between the houses. Some of the houses even had small gardens. There were also wildlife sanctuaries with small lakes stocked with fish. Water had become more available for such wildlife ventures and people were adding it to their home property to avoid the former encroachment upon the wildlife. We had those generations of giving our children a proper education at all levels. The large scale molecular manufacturing was still not a reality, but it could come into the scene any day now and people would be ready for it.

There was a brisk wind and numerous wind mills turned rapidly and some made slight sounds. I walked west on Main Street, away from the intersection. I entertained the idea of calling the Vansar Computer Store. I'd have to use my cell to call the store on Route 80. I continued my walk, glancing back at the convenience store where I was parked, now and then. When I was about a hundred yards away, I stopped under the shade of a stately elm tree and I took the cell phone out of my pocket and punched in the number for the store that was on the literature I had taken from there. The voice that answered I recognized as Jim's, the man Torres had turned over to our side.

"Jim, would you call Captain Torres to the phone?" I asked. He knew my voice too.

"This is Joe Torres."

"I can see the house from where I am now," I said. "I'm watching it while I talk."

"That's good, any activity?"

"The wind mills are turning like crazy."

"Nobody called here while you were driving over there either.

"I'm going a little closer to look around," I said. "Will you be there a while?"

"Yes, till my man gets here."

"You've got my cell number."

"I've got it, Dan, so stay in touch," he said. "My man just came in the door."

We ended our call and I moved closer to the house, playing the casual walker who needed the exercise. This was a well built fieldstone house with the number 2531 on it, the only house that was visible from the intersection of Lockhart and Main Streets. Joe was right about it. He remembered driving by there and admiring the fieldstone siding. The number on the carved cedar sign matched the number he gave me. I walked on past the front view of it on Main Street to a two lane private road marked Dewey Lane. I turned left onto it and went on the sidewalk that ran along to give a right side view of the house. I kept walking till I saw a fieldstone Garage in back of the house and I went to the other side of the lane to the adequate driveway between the house and garage. I could hear the sound of a power tool being used in the garage and I walked near enough to see inside, where a man was bending over a work-bench. He looked up and his face was familiar so I walked to the garage entrance.

"I know you from Comp-ease," I said.

"Yeah, I remember you, Dan Brandon," he said. "I'm shop foreman, Jeb Smith."

"Yes, that was where we met there in the shop. It's changed a lot since then."

"You couldn't just walk in there now like you did," he said. "How did you happen to be out this way?"

"I stopped at the convenience store on Lockhart Street and just started walking the way I often do. On my job I'm in my car driving someplace too much and don't get enough exercise."

"I'll bet you do a lot of networking, too, that way."

"Yes, my share of it," I said. "I see you have your own shop here." I walked a little closer and saw he was putting together a computer. He noticed my look of surprise when I saw the small assembly line on the bench with a complete Vansar at the end of it.

"It's moonlighting," he said. "I responded to an advertisement in the paper for small contractors to put together some computers on their own time."

"Oh, I see it's a Vansar," I said. "I think I've seen that brand before."

"Where have you seen them?" He asked.

"I'm not sure, but in a store somewhere around here." I glanced back at the top of his work-bench and estimated that there were the makings of two more besides the one he had already assembled.

"Well, that's news to me," Jeb said. "I haven't seen them around town."

Something here wasn't adding up. Jim had the number he had been given and it turned out to be the phone listed for this house. And yet my friend Jeb from Comp-ease acted like just another employee, not the boss. I didn't think he was telling the truth.

"It's getting dark already," I said. "I guess I've walked to far away from my car and I better get headed back."

"Jump in the pick-up," he said. "I'll drop you off at the store. You're a long way from there."

I was glad to do as he said and it only took a few minutes for him to drive me to my car. I told him that I did appreciate the lift. He smiled and waved as he drove away, like an old buddy of mine. I pulled out my cell and called Lebeau's cell number and he answered on the first ring.

"They say the boss is the last to leave," I said.

"Yes, I recognize your voice, Dan, thought you'd call late so I left the cell open."

"You're on the way home, so I'll be brief. Captain Torres has a man watching the Vansar store. The man at the store said he met the boss once and was given a phone number to call to turn in orders for more computers. Captain Torres had the number traced to a house in Martindale. I went to that address and found Jeb Smith was in his shop turning out a few Vansars. We were both surprised to see each other."

"Someone's just using him," Lebeau said. "He's only a technician not an engineer, and the person running a new start up for molecular manufacturing must have a Doctorate Degree in electrical engineering and nanotechnology, like I do."

"Oh, I get it now," I said. "So you're the one with the Doctorate at Comp-ease, but you have a research scientist like Bob Clauson, who has the latest expertise you need. It was the same way at Gladstone."

"Smith makes a good shop foreman, but he can't be running the show at Vansar."

"He told me he saw the job advertised in the newspaper under small contractors."

"Sure, we don't care if he does that on his own time, "Lebeau said. "Keep looking for the one that's writing checks or maybe even paying in cash."

"I'll keep after it," I said. "I'm getting good help from Captain Torres."

We closed our cells and I got out of the car and went into the side door of the convenience store. I bought a tuna sandwich, soft drink, and a newspaper to take back to my motel room. Climbing back in the car, I drove onto Route 80 and back to Luling.

As I parked in a space next to my room, I heard Willie's voice coming from several cars away. He said that he had to have one more report from me, before he went back to San Antonio. We went over to my room, where I still had a couple extra chairs and we got comfortable.

"So, Ray went back to San Antonio?" I asked.

"No, He had a date with Chris, there in Crystalville. I told them he ought to take her back to San Antonio with him, plenty of cop jobs there."

"That's right, be easier on both of them. I'm glad they got together, however it turns out."

"She's taking a few days of her vacation time, and I told him he could always go back with me, after he took off as much as he wanted. Yeah, they're getting it on."

"Did Harold, settle up with you guys right?"

"I couldn't believe how big the checks were, and naturally we compared checks. They paid all our medical bills and plenty of incidentals I never even thought of. We got a nice bonus too. Harold said you would get one."

"It's good to know all this stuff, I'm glad you stopped by on your way home to tell me about it."

"You're not getting off that easy," Willie said. "I want to know what you've been getting yourself into while I went to Crystalville and back. By the way, Marsha asked about you, and she gave me the latest poop at Gladstone."

"I'm going to be fine," I said. "This is starting to look more like a normal case and I even have Joe Torres assigning men to help me with things in the Luling area.

"That makes me feel better, so that's really good that it's getting like a normal case."

When I asked him what Marsha had to say, Willie gave me that wise-guy look and I knew he had appraised my relationship with Marsha for just about what it had been. I took the time to clear it all up with him. He had become aware that Marsha was a pretty fine person, and I assured him that she knew about Claire and there wasn't a problem. That was what I said anyway.

The latest thing at Gladstone was that Nelda Whittington had just completed the rest of her work to receive a Doctorate from the University. I wondered about how she would get along with John Von Mueller. Sometimes two Doctorates in the same company would be one too many. Willie seemed to think they were getting along very well. Then I remembered that it was Von who had gone over to the University and brought back Nelda to start her job at Gladstone. She must have already done most of her graduate work.

"According to Carl Watkins, Remkin was the main problem at Gladstone," I said. "How is the new cyborg security system going in there?"

"Remkin and Carl are a team," Willie said. "There's a cyborg like Steve Wingate who's there full time to work with security. I didn't meet him but Carl told me about it. Carl said if he gets something he can't handle, the cyborg can. It's a great organization now."

I thought Willie was like me, in that he took a personal interest in the people who were clients. Some investigators I had known were highly intelligent, but too schizoid to get the job done. Willie and I could talk all night about the people we knew on various cases, going from one catharsis to the next one by being reminded of a client. Now, here he was with a big paycheck, and a wife, who loved him waiting for him at home. I didn't need to be encouraging this conversation. I thought that I would just shut up and he'd go home to Helen. It took more than that. He enjoyed leading me through more of the past escapades and I was reduced to an occasional nod or a stifled laugh. I finally yawned and rubbed my eyes.

"I'm getting sleepy Willie," I said.

"Yeah, I better let you get some rest. He yawned like it was catching. We shook on it and he was out the door.

Remembering that I had turned on the TV to watch the news, I wakened with a stiff neck. The TV had switched itself off after a while, and helped me to get into a deeper sleep. I got up, went to bed right, and slept a few more hours before rising.

CHAPTER 40

It was a beautiful sunny morning and I felt more like going on a picnic or writing a poem, but I made myself get back on the horse. In my case the horse was a used version of one of those non-polluting, energy saving cars everybody had these days. I drove over to the usual location where I parked and walked to the hilltop surveillance point for Clauson's safe-house. My pocket binoculars proved adequate for me to have a closer look at anything in this area I wanted to see. About two hundred yards north of the house I saw several young children playing in their yards, while their young mothers talked across the fences. The birds chirped so loud in this sanctuary that I couldn't hear anything else, over their emphatic remarks to each other.

Turning my attention closer to the safe-house, I scanned in front, to the sides, and behind it. Then, I adjusted the focal point and started viewing inside of the windows. That was when I saw a hand from the outer region of my image reach into a basket of fruit and take out an apple. The hand with the apple seemed to disappear as soon as I had seen it. I took my cell phone out of my pocket and pecked out the number for Steve Wingate.

"There's someone in Bob's house," I said.

"Oh, I was just getting ready to call you," Steve said. "Virgil, is going to meet him there for his therapy session this morning. Virgil's a tall sinewy man with glasses and ample nose to hold them. He's always clean shaven to dispel the image of the frosty bearded psychiatrist, but he does have long blond hair."

"Thanks, and I'll know not to bother them, now."

"Yes, thanks to you for calling," he said. "Sorry I can't chat any longer, we're trying a new nanotech thing at the factory and I have to go now."

It was a trade secret, so he couldn't talk about it. I wished him good luck with the new attempt and we closed our cells. As I fiddled, putting my cell away it occurred to me that something didn't ring true. Bob had an appointment with the Psychiatrist. Something new was going on at Comp-ease. It looked to me like Bob should be there with Steve.

I went back to watching the neighborhood. I heard something, a rabbit or armadillo rushing through the underbrush, and I thought the birds chirped even louder. Some of the birds flew away and I knew something was afoot. I picked up the poncho I had been sitting on and the field-glasses. I went to the edge of the thicket. I ducked behind the trunks of several large trees, where I could still watch the spot I had retreated from. In only a few minutes a man and a woman emerged from the trail and walked over to my spot at the top of the hill. They both wore pants and sport shirts. The woman was the aggressor and walked right away to where I had the poncho on the ground. She kicked around half pointing her foot at the spots on the ground where the edges of the poncho had left tell-tale marks that revealed cleaner ground in the area that had been under the poncho. They looked down at the safe-house, then turned around and went back down the trail out of sight.

Following a few seconds behind them, I saw them go out into the road and walk about a half mile down it. They got into a waiting car with an engine running, and it drove farther down the road in the direction of, where I parked my car. I couldn't make out what kind of car it was, but I thought it might have been one of the cars belonging to the new CIA people guarding the general whereabouts of Bob Clauson. I had forgotten all about those three newcomers. They had all been formally dressed when I had seen them at the restaurant that day.

Especially, the woman looked different today, in her more masculine clothing.

I went back up to my roost and watched till the psychiatrist parked in front of the house and went inside. In about an hour the man came back outside and drove away. I thought that would be enough observation for a while. Moving down the trail to the tent, I examined it and found nothing in it had been tampered with. Nevertheless, I gathered together some of the things in a plastic bag and walked the rest of the way to the bottom of the hill and down the road to my car.

Throwing the bag inside on the seat, I drove back to the place at the foot of the trail.

I parked the car, opened the trunk and tossing the bag inside, I hiked the short distance to the tent. It didn't take much time to pull up the stakes and roll up the tent. I folded up the cot and put it on one shoulder, the rolled up tent on the other one, and walked back to the car with them. I tossed them in the trunk and climbed back in the car. I wouldn't be spending so much time on the hill now.

There was a nice breeze blowing through the trees and I put down the windows to enjoy it. I still had to work on this beautiful day, but I didn't need to get any farther removed from it than I had to. Since we had joined the rest of the world to stop global warming and renew the world, we had started riding bicycles, using solar cells and all the right things. That had been the turning point and America regained its stature, finding the courage to start doing the right things. We got rid of ungrounded fears about evil terrorists coming here after us, if we didn't go to their home and get them first.

Thinking I needed to keep up with things going on locally, I drove to the east side of the river, came into Luling from the south side of Route 80 to the new industrial center that had begun with Comp-ease. The Gemini people were moving along at a surprising rate for a company that had opened its doors only a few days ago. These computers were essentially built for Space Age agencies like NASA to fit inside of a spacecraft. As I went inside the building, the sounds and odors of construction met me. My first impression was that I had been there before. In a few moments I knew that it was the entrance, which had been reproduced exactly like the entrance at Comp-ease. I scanned the printed instructions on the counter top, pressed the button and had a seat.

In a few minutes an attractive blonde woman came to the counter, smiled and said she was a temporary receptionist. I asked if I could see John Chin. She pushed the lever and opened the door to let me go over to her side. Leading me down the corridor to the last office, she had me wait at the door. John Chin came with an outstretched hand and we shook. He ushered me inside to sit on the other side of a large mahogany desk.

"Good to see you again, Mr. Brandon," Chin said. "Can I be of some help to you?"

"Yes, and there are times when I can help you. Some of the same things regarding general security, trade secrets, and ethical business practices that are problems at your company will also be problems at Comp-ease and at Gladstone. I've been trying to spread the word about the problems I'm working on. Sometimes we can help each other a lot by just keeping in touch. Let me give you an example by telling about James Lebeau at Comp-ease. He told me he has another engineer friend at Aster Computers in Lockhart and they talk on their cell phones every day about things they have in common. This engineer at Aster told James about a new computer company on Route 80 near Martindale. They both suspected some kind of fraud was going on there and I went to check it out with Captain Torres from the Luling Police. Joe Torres put a man over there to watch the place for a while. When we satisfy ourselves that it is fraud I'll warn you about it. If you hear anything that sounds suspicious give me a call on the cell phone. I'll check it out and get back to you. If we work together, maybe we can stay ahead of things."

"Yes, please count me in," Chin said. "If I hear of anything like that I'll call you right away, and about this new company, anything we should know about it to protect ourselves?"

"The company name is Vansar, so if anyone should approach you from there I suggest you don't do any business with them, and call Captain Torres as soon as they're gone."

"Wouldn't it be too late for the police to get them, when they're gone?"

"We don't want you putting yourself in harms way," I said. "You could write down a few notes of description or a license tag number without them knowing about it."

"Oh, I understand, that is all very helpful," he said. "Won't there be a charge for the service?"

"No, I'm already well paid by one company and another one has offered to pick me up if that one terminates my services. I'm very secure in the security business." We both laughed at our bit of an inside joke.

As I was leaving I could still hear John Chin's chuckle coming from his office. He was a highly intelligent man who appreciated everything anyone did for him, and always wanted to do his part as well. The same young woman who said she was a temporary receptionist followed me back to the front and opened the gate for me. I went outside to my car. Since I was so close to Joe Torres's office I drove a few more blocks north on Route 80 and parked in front of the LPD building. I went inside and found Joe in his office.

I told Joe that I was instructing the CEO's of companies like Gemini, to let us know if they were contacted by anybody from Vansar.

"That just might do us some good." He said. "Anything come out of you checking that place on Lockhart Street?"

"It turned out to be a shop foreman from Comp-ease who took on the job through a newspaper advertisement. That number he called is no longer in service, as you might have imagined."

"So, the man from Comp-ease just took it on to make a few extra bucks," Joe said.

"Lebeau said they don't mind if their people moonlight on the side," I said. "The foreman, Jeb Smith, is just being used as a technician to put a few computers together. He said he was sure Jeb was not involved in any more than that, and I have to agree with him."

"Do you think the slippery character running things, already knew the man from Comp-ease and ran the ad in the paper to avoid personal contact?" Joe asked.

"Good chance of it," I said. "He seems to need to be anonymous."

"We thought that was because he was planning to do something illegal, but maybe it's more than that. What if he has a somewhat well-known face?"

When Joe said that, it got me thinking of all these people who were working in the electronics field, and about who could work out such a clandestine program of fraud. It was the largest technological scheme I could think of. The perpetrator would be on a par with Albert Einstein, to me, since it involved the far out concept of molecular manufacturing. It would have to be someone with an impressive Doctorate who would be able to walk into a computer company, and convince the top brass to pay him to buy into the new nanotechnology game. It had to be the biggest game in town. I could understand how some of the super-computer companies who didn't have it, might want to invest in it. They'd think the investment of a large amount of money would be well worth it, because molecular manufacturing was going to allow them to produce a lot more products in a much shorter time than the old traditional manufacturing process. In a very short period of time they would get back the money they paid out to get into the nanotechnology and molecular manufacturing. After that initial investment they would be into the new age of manufacturing. If they didn't make the investment they would be missing out on the New Industrial Revolution.

CHAPTER 41

It had been a gradual progression for me to become somebody, who was able to have empathy for those people in the upper echelon of industrial technology. These individuals were playing for big stakes in the nanotechnology game. If their company didn't keep up they could lose everything they had, and the higher up you were the farther you would fall. The people like me would just pick themselves up and go get another job, but it wasn't so easy for those at the top. Sometimes there wasn't even another job available for them. I didn't think there would be many job openings for a person with a Doctorate in Electrical Engineering. All of those computer companies probably had one or two of them, but there were a lot of companies.

Before I left Joe Torres in his office, I told him about my general plan to try to enlist the help of Jeb Smith in the same way we had done with the man at the Vansar Computer Store. Joe said he would drive his LPD van to Comp-ease and meet me there. It was early afternoon when I parked in a visitor's space and got out. Joe's van was already parked there and we went inside together.

Since Jeb was the shop foreman, I thought I might get away with just asking to speak to him for a few minutes. It wouldn't take longer than that, once Joe flipped the badge at him. It did work, as Jeb came along very quickly, as though he might have been told they could only spare him for a few minutes.

"This is Captain Torres, LPD, Jeb," I said. "He'd like to talk to you a few minutes about your part time job, putting together a few of those Vansar Computers." They shook and Torres flipped open his wallet for a second or two.

"Is there something illegal about that company?" Jeb asked.

"It's what we call a front company," Torres said. "The computers you're making to put in the store are perfectly legal and the store's legal, but it's only a front. What your boss wants to do is take orders for more powerful, smaller computers."

"I don't know the boss, never met him."

"That's OK, Jeb," Torres said. "He'll probably send James, the store man around to you with the money. You get paid and it's legal. What we need to do is find out who the boss is, and if you could just hang in there for a while I'm sure we can do it and I promise you that you won't be in any trouble."

"What did you want me to do?"

"Just let me put somebody with you at the time you start your evening job we need to tap into your phone and be ready if this person calls you. We could trace the call and find exactly where he's calling from and go after him."

"Well, I guess I should, since I let myself get involved."

"Not your fault, no way you could know. What time do you get home to your shop?"

"Six o'clock."

"There'll be an officer there in plain clothes, nothing to identify police. I'll call you first to verify the name of the officer who'll work with you."

Jeb Smith agreed to cooperate with the LPD, and then hurried back to the shop. Joe and I went back outside to our cars. I told Joe I was going to start investigating a few of the people I knew who might qualify as a nanotech person, with skills that could put a molecular manufacturing machine together for a computer company. Joe drove off in his van while I rousted around a few minutes in my office in the car. I came up with the home address and phone number for Dr. Wallace Norton. He lived in Hyker Park, a convenient place to live near his job at the University.

In about a half hour I was getting off Highway 183 in Austin. I drove to the campus location of the yellow painted saw horses and blinking red safety lights. I parked at the hole in the wall place and walked into the tunnel. When I came to the locked door at the other end of the tunnel, I pounded on the door and waited. An average sized young man with wire framed glasses unlocked the door and said, "This is the electronics research lab, that what you're looking for?"

"Yes, it is," I said. "I'm Dan Brandon, used to work with Nelda Whittington here."

"I'm Donald Henson," He said. "Nelda's working at Gladstone Electronics now."

I handed Donald my business card and he looked at it and then motioned me inside.

"This is just one of those break rooms for all of us working in this area," he said. "Come on inside, if you like, and have a sit down."

"Thank you, I think I have the change for a drink and a sandwich."

"Dollar bills will work too," he said as he carried his coke to a table and put it down. I took my tuna salad sandwich and a coke to the same table and put it down.

"I'll have that sit down now," I said.

"It was my Dad that used to say, 'have a sit down' and that's how I got the habit."

"My Dad used to say, 'finer than frog legs split nine times'," I said.

"That's a good one too. I guess we're under investigation."

"Well, that's what I do.

"I'm ready to spill my guts any time."

"Are you a Dr. Donald or close to being one?"

"I'm not even close to one," he said. "Does that mean I'm not a suspect?"

"You're not a good suspect."

"Would Dr. Wallace Norton be a better suspect?"

"Yes, he'd be considerably better."

"Wallace is onto you too," Donald said. "He called you a PI checker outer"

"Where is he now?"

"He's in the lab, while I'm taking a break." He laughed for the first time.

I nibbled at my sandwich and took a big swig of coke. "Well, when you're finished your break, would you tell him to come out here so I can interrogate him?"

"Great, I'd love for you to do that. Those PhD guys have been giving me fits."

"It's just Piled Higher and Deeper," I said.

"Amen to that." He said.

"Bob Clauson didn't like them so well either."

"Nope, we used to cut them all the time."

"Don't forget to tell Wallace I'm here to harass him."

Donald laughed uproariously at that, tossed his coke can in the trash can and went in the lab. He closed and locked the door behind himself, to execute the proper procedure. In a few more minutes, Wallace Norton opened the lab door and locked it after himself.

"Donald said you're here to give me a thorough interrogation," he said smiling.

"You do know that when I come around I'm looking for some answers."

"Sure I do, it's your job to get answers."

"Did your Doctorate Degree include work in nanotechnology that would allow you to build a molecular manufacturing machine?"

"No, it did not," He said. "It was in electrical engineering with emphasis on methods of research."

"That's the name of the course you teach Methods of Research?"

"Yes there's a great deal of preparation needed to teach it."

"What sort of research methods are you teaching them?" I asked

"Only the general research, they can apply to their own fields. They pick a title and type up a practice thesis. I have to grade all of them and give a mid-term exam and a final exam."

"Do you get a chance to leave campus to do some shopping?"

"No, I never leave campus. My wife has her own car and does all the shopping."

"Sounds like they keep you busy over here."

"What I appreciate is your direct approach, Mr. Brandon," he said. "It allows me to understand what you're driving at and possibly even be of some help. On this occasion it seems that one of us has been engaged in some sort of extra curricular activity that is not ethical or maybe even illegal. That lets me out—don't know or care about nanometers."

"I'm beginning to think whoever it is, must have a lot of freedom to travel around, and yes, it seems that I can eliminate you. That's all I can say about it, Mr. Norton."

"I understand, so say no more, feel free to come by any time you need help."

Tossing my trash in the can I went back through the tunnel to my car, and I was fortunate enough not to have a ticket on it. I didn't tell Norton that a suspect was never really off my list till I was off of that case. It meant I kept an open mind. Even though it might seem impossible for someone to be a perpetrator, where there's a will there's a way and I've found anything is possible. Unfortunately that means I'm working right through to the end of many cases, with most of the suspects still in place when I solve the case. And in the end I might put one or two people in jail, hoping they're the right people.

I had left Luling to go to Austin and have a face to face with Wallace Norton. Why would any of those who were already into nanotechnology and molecular manufacturing, be out around central Texas trying to get set up into something they were already into, and doing well at? I found it left Norton out, because he couldn't get away from his classroom long enough to go to the grocery.

Getting back in my car again, I drove onto LH-1 and headed north to Crystalville and I suppose it might get all out of kilter back in Luling.

CHAPTER 42

The polished granite of the Gladstone building sparkled up ahead of me on LH-1 and I turned off on the access road. I thought the shining granite had some kind of magnetic pull that made me keep coming back to it. I drove off the access road to the last loop of private road that took me to the parking lot on the left side of the building. The parking lot was full, and I had to circle around several times before I could see a visitor's spot open. I parked in it and walked to the new entrance section, which had been constructed like the other security systems built for the Luling computer companies. I suspected there were no more than five start ups Clauson was working with now, and it would be as easy as walking through the front door to identify them with the same special cyborg security set up.

Pushing the easy button, I didn't even have time to sit down in the small reception area, before Carl appeared and opened the gate for me. We walked down the short corridor that let to the old familiar section of the first floor where I could see the elevator on the left and straight ahead of us was Carl's new office, built against the far wall where there used to be chairs for small group meetings. They were gone and Carl's office occupied the space, but there were several plush chairs in the office, and a good sized coffee urn. He responded to the changes by becoming a confident, happy man.

"Sit a spell Brandon, I've got someone I want you to meet," he said. I noticed Carl had lost his suit and tie. He was dressed casually now. I wondered if he even bothered, any more, to flash his deputy sheriff badge on anyone.

When Carl returned he had a tall, casually dressed man with him. I saw the man had a left hand that was robotic.

"Dan Brandon, this is Edmond Tomlin," Carl said. "He graduated from the Cybernetics Institute of New Hampshire."

Edmond tossed his right hand at me and we shook. I wondered if he was really left handed. "Carl's mentioned, what a help you've been, Mr. Brandon," he said.

"All I did was hook him up with the right people, just call me Dan."

"Ed's what they call me."

"He did three tours in Iraq, before he went to the Cybernetics Institute," Carl said. "Let's have some coffee." He drew his own in one of the plastic cups, and Ed and I did the same and we all sat down in one of the soft cushioned chairs with an end table to place the cups on if we chose.

"Have some donuts too," Carl said. There was a box of them, with napkins, in front of us. I was already amazed at how well Ed used the robotic hand to fix his coffee with cream and sugar, but when he finished eating a donut, he took out a tooth pick and cleaned his teeth using the robotic hand to hold the tooth pick.

"Well, I'd better get back," Ed said. "Glad to have met you, Dan."

"Glad to meet you too, Ed," I said. He went back the way they just came, and I asked Carl no questions.

"Sure, I've got the title, head of security, but Ed really handles the job by himself."

"Don't sell your self short, Carl," I said. "I know you had a job as chief of police in a large city, when you retired and came here. That kind of experience is worth a lot."

"Oh, yeah, I'm not worried about keeping my job, but Ed's a real hero."

"I think that way too," I said. "I did a tour in Nam, but I'm no hero. If a guy only does one tour and comes back missing any part of his body, that's a hero to me."

"He lost the hand in Iraq, but the robotic one was refined some at the Cybernetics Institute," Carl said. "He told me he can feel the robotic fingers through the neurons to his brain just like he does with his real hand. He had a computer chip brain transplant too."

When I told Carl about the article I had read about how people with robotic parts of their body could have empathy with a robot, he said he had heard that. We both had grown to have respect for people who had undergone such things. We didn't want to be cyborgs ourselves, but we could appreciate the need for them. I remembered how Willie had remarked that Steve Wingate seemed to him to be extra sensory.

"What did you think of the way Edmond handled the toothpick with the robotic hand?" Carl asked.

"I've never seen anything like that before,"

"Not to change the subject," he said. "But I know you're here for something more than just to be social."

"I'm especially interested in the recent activities of those with a Doctorate in Electrical Engineering," I said. "I know Nelda Whittington just got hers and that makes two of them here."

"What sort of activities are we talking about?"

"I'm talking about selling their own specialty to other computer companies."

"Since both of them work for this company it'd be unethical, but I don't think illegal."

Changing the subject slightly to which one of the two would have the better opportunity, I gathered that Carl thought Von Mueller would. He was out and around doing speaking engagements and that sort of thing, when Nelda was here at her desk on the first floor, not far from Carl, every single day working hard and at top speed. Von had just about made up his own hours and nobody seemed to know what they were.

"Wouldn't Remkin know when Von Mueller was supposed to be working?" I asked. "Haven't you ever asked Remkin about Von Mueller?"

"I've asked him plenty of times," he said "He always said the same sort of thing. It was always a conference, being a guest speaker or some such thing, so I got tired of asking him. It's either the old anything to back up another administrator thing, blackmail, or some combination of the two."

It got me started thinking about the relationship between Remkin and Von Mueller as I had observed it. I was taken back to a certain day, when I discovered that there was an extra key to the research lab where the super-laptop had been kept. It was Remkin who told me that, and he showed me the place where they were supposed to be kept in his office. That was when we found one of them was missing. He had been very upset, and when I suggested Von Mueller might have it he claimed Von Mueller never had access to the vault where the keys were kept. Remkin was supposed to be the only one who knew the combination. I doubted that Von wouldn't have known it then, and now I doubted it even more. In those days I trusted all the administrators. A lot of things have happened since then.

"Did you ever hear Remkin and Von talking about the keys to the research lab?" I asked.

"No they never talked about it in front of me. I didn't want to do any more than take a key from Remkin to go in there and then take it back to him."

"Did you know that Remkin had an extra key and somehow lost it?"

"No, I didn't know anything about that."

"Well, if someone had the extra key they could have gone in the lab early that morning and taken the laptop," I said. "And then Remkin unlocked it with his key, went in and discovered the laptop was gone."

I thought that was exactly what had happened. Carl had to admit that security in those days was not so good, although he was not directly to blame for it. I didn't believe George Campos, who had been the afternoon guard, had anything to do with the computer theft. It seemed likely that Campos, Gilbert, and Oliver did nothing more than rough up Clauson and hold him somewhere for a while. They expected to get something out of it, but what they got was killed over it. They had never had possession of the super-laptop.

It wasn't fair to expect Carl to know where Von Mueller was, when the man obviously didn't want to be accountable. Carl was due to make his rounds through the entire three floors of the building. I wanted to talk to Nelda since she was only in the next office to the left. I told Carl I knew how to open the gate and let myself out. When I glanced in Nelda's office she was on the phone and she motioned for me to come inside and be seated next to her. She looked better than ever. Since I had known her she had gone from short, straight, sandy-colored hair to a beautiful blonde done up professionally high on the head.

"That was Marsha," Nelda said. "Since there wasn't room for me up there with the rest of them, we have to do a lot over the cells to save steps back and forth all day."

"Have you thought of changing offices with John Von Mueller?" I said."

"No, I haven't thought of that," she smiled. "Should I?"

"You could ask him about it, nothing to loose. He's just in and out anyway, might like it better here on the ground floor."

"Hmm, don't think so, as Von's conscious of appearances. He could get mad."

"Just a thought," I said. "Do you know where he is today?"

"Oh, well no, let me tell you how it is, Dan," she smiled. "Von did really run things here for a while, and then he had the chance to really be the CEO, if Harold would have taken his retirement. Relations between them disintegrated from that time on. That was when Von came up with the idea that he was worth more to the company on the road, making speaking engagements, than he was hanging around here.

And Harold started being really obnoxious to everyone, so we all tried to stay away from him and that's what he wanted."

The message box finally penetrated through my thick skull and on into the brain. It all fitted together in the box and made a lot of sense, and it was all about Von. I could have been traveling all over this big state looking for the Vansar molecular manufacturing man and never found him. I had to come here and have a one on one to get it. What Von was doing was using his name from having been the main speaker at the Global Technology Conference. There was a list of participants that he had to go by. They had all registered as participants giving names, addresses, phone numbers, and the name of the company they worked for. He could have it set up to go to a different company every day to give his nanotech speech and go through the molecular manufacturing startup pitch at the same time. He would be getting some takers and if he got enough of them, each with a nice check made out to Vansar Electronics he might be about ready to quit at Gladstone and go into his new business full time.

"Can you remember anyone mentioning one place where he had a speech engagement in the last month or two, Nelda?" I asked. Thinking I would be able to take the list of contributors I had kept and estimate about where he might be going next. If it was narrowed down I could even call a few companies and ask when he was scheduled to come.

"No, I really haven't," she said. "I told you how it is and I just come in every day, do my own work and don't worry about anybody else at Gladstone."

"I had no idea it was so hard to work here, Nelda, Can I do anything to help?"

"Yes, tell me how Bob's doing down there in Luling?"

"I enjoyed having lunch with him a few days ago in Luling. He seemed to be well."

"He seems to like it at Comp-ease, but things can change so quickly," she said. "Not long ago I wanted him to come back to Gladstone and now I'm glad he didn't, I'd just like for us to be able to work in the same town some day." Her voice broke.

"Just hang in there and it's going to work out for you two," I said. "I won't give you any more advice about Gladstone, since you know a lot more about it than I do."

"That's what most of my friends have been saying, that it takes time. You and Carl have been good friends and I want you to know I appreciate it."

"Yes, I found Carl to be a loyal friend after we got to know each other," I said. "I'd like you to just forget about me trying to run down Von Mueller, and I won't ask you any more questions."

"He's a shark like Remkin. I won't tell either one of them anything about you."

Assuring Nelda, as one of my best friends, I told her that I was also careful not to repeat anything about my friends that could cause any problems. I said I'd tell Bob about our nice visit as soon as I got back to Luling. She smiled at that and I went back outside to my car.

CHAPTER 43

It was a good thing I had developed the office in the car concept. Discovering that as I moved around on the job, I was always wishing I had something with me I had forgotten, I came up with a few portable storage compartments in the front of my car. Small items like business cards I kept in small boxes in the glove compartment. The rest of it consisted of the recent file folders I kept in a leather case that I could toss on the floor or on the front seat.

I found the list of contributors to the Global Technology Conference and was glad to see the list was alphabetical by the name and address of the company. Scanning the list I could easily pick out the ones in Texas. At that time I remembered how many more electronics companies had been added in my generation. But still, I was surprised there were so many starting with the letter A. I remembered Aster Computers in Lockhart where Lebeau's friend was an electrical engineer. There were a lot of names that came before Aster. If Von had already gone to speak at every company starting with Abco Electronics and ending with Aster I would have been amazed. He could be skipping around. If he scanned the list looking at the addresses, he'd probably mark the ones closest to Crystalville and visit those first. I thought that would be most likely.

It came to me, as I glanced at my handy eight by ten map of South Central Texas. Maybe I was still going about it in the wrong way. I thought Von might have decided to just go to the largest city in his area, where there were more electronics companies than anywhere else. It was only twenty minutes away, and he could set up all his appointments by networking. I thought that was the best way. I could go to Austin, do the same thing and I'd cross his tracks. I put the map back in the folder.

Starting the engine, I drove to the south lanes of LH-1 and put my foot to the floor. It was great not to be using gasoline any more. I was old enough to be able to remember those days. Now you had no sulfur-like odors or air-pollution. You could put your foot down without giving yourself a guilt complex.

Making my usual transition from LH-1 to North Lamar, I drove south to MLK and then east to the drag. If I ran into Von there were about a thousand reasons I could be in Austin and he wouldn't ask anyway. I even had a great reason to attend his recent speaking engagements in Austin. It was a sorry day, when I had to miss his presentation at the Global Technology Conference. I could feel so guilty that I would attend his speeches in Austin over and over till I got it right.

There had always been plenty of oil money at the University just like it had been at Luling. Those people can take the dozers to half of their city if they want to, and move things around like chess pieces on a chess board, till they get the right combinations to go on making money. At the time when the computer industry just exploded about two decades ago, The University liked them and those companies started to move into an area where the drag went north away from campus. And then before long the University bought more land and expanded the entire area. It was full of high class electronic businesses who wanted to be in the university atmosphere. Most of them even had their own medium sized auditoriums that were in continuous use.

It wasn't much of a surprise to me, when I came to, Beaker Electronics, where I found Von Mueller had given his presentation. I made note of the date he made it and the title. The title had been Modern Nanotechnology and Molecular Techniques. The woman I talked to was an electrical engineer, and I think she thought I was one too. I wanted to know if Beaker had signed a check, and signed up Von to be their ticket to molecular manufacturing, but I couldn't come right out and ask.

"That Von's really into molecular," I said. "Everybody's going to have it sooner or later." Her eyes lit up and she seemed impressed.

"It's going to be sooner for us," she said.

"Yes, that's good." I said. Waving to her with a big smile, I worked my way to the door and outside, before I could be questioned. I got to the car in a hurry and on to the next place. My intention was to document all of Von Mueller's recent speaking engagements and I doubted I'd need to leave Austin to do it.

The layout of the north drag electronics companies was for one way traffic. There was a separation island of concrete between the buildings. You turned off to the right to go to each company on short loops that routed you to that property. When you reached the end of the high-tech industrial park, the road turned completely around and you'd drive back the way you came and have about the same number of companies to visit on your way out. There was a directory of companies as you entered the compound and there were also individual signs for the companies as you came to them.

Beaker had been the third company I had come to and I had success. The first two had not been contacted by Von Mueller for some reason. All I could do was take them as I came to them and I thought it was a good system. The next company, Janco Computers, had him scheduled to speak in thirty minutes with a title of Modern Nanotechnology and the New Industrial Revolution. I drove up a slight incline on the loop and came to the auditorium. It came complete with a small parking lot. The parking lot was almost full when I pulled into one of the spaces and sat there for a few minutes to relax and get my portable office straightened up. I folded a few sheets of typing paper and put it in my pocket, in case I wanted to document something molecular he had in mind for the revolution. I didn't think he was going to do anything illegal here at Janco, but if I could connect it to those Vansar Computers it might be fraud.

Drifting inside the room that was a lot like the small individual theaters in a local shopping Mall, I sat down in a chair in one of the last few rows of empty chairs. There was a good PA system so I could hear as well as somebody in the front row. Von was chatting with a few of them in the front, when I had drifted in with the other stragglers. He was trying to establish some sort of rapport with them, that might spread to other rows if he did it right. He had been talking about the good job the women's basketball coach at UT had done with character-building. There were a lot of women in the audience.

"Well, I guess it's time to start this thing." He said.

It got quiet without a problem. A few in front were still talking about women's basketball, but he just smiled and waited for them to quit and they did.

"Nanotechnology is now established as a fact of life," he said. "I'm John Von Mueller and I recently earned a Doctorate in Electrical Engineering at UT. I've been working for Gladstone Electronics. I can assure you that at Gladstone we believe in every aspect of nanotechnology. And you can tell it by looking at our small laptop computers."

Von talked about nanotechnology for close to an hour, and I was beginning to think I even understood it. He was aware of another fact of life, the one about how nobody could talk for more than an hour with the same enthusiasm that charmed an audience. He had set the hook and was ready to reel them in.

"Some of you probably know how we have carried it to the highest level," He said. "We have taken it into molecular manufacturing, and at Gladstone they are actually selling shares into it . . ."

"Isn't that very expensive?" Someone asked. I wondered if it was a plant.

"Yes, you have to be another computer company with a high level of accomplishment in the small, super-computer realm, and can be set up for grants in addition to what Gladstone does for you. You have to go over there to learn about that, I don't handle it."

"How could an individual get into it?" Another asked.

"If some of you want to stay and talk about that I've established a way," he said. "I offer myself as a consultant at a reasonable fee, nothing like what you pay for an entire company."

Almost half of the crowd was interested and gathered around him in a circle. I didn't need to stay, but I hung back enough to see he was having them all fill out cards with names and phone numbers.

"I had no idea there would be so many of you," he said. "I have some free literature to give out on the subject. You'd need to read it first to go to the next step. It's programmed learning. I could start a few of you on it now, but the others would be standing around waiting. It would be much better for those of you who could schedule to have your own individual appointments within the next few days. If you give me the cards with your contact information, I promise to call you back." Over half of this group turned in the cards to him and left. If only one or two in the call back group gave him a sizeable check to get in on the ground floor of molecular manufacturing, he would be in business. It seemed like he had been spending most of his time setting up programmed information and getting special nanotech computers ready to go. He was working with a select group of engineers who were close to getting doctorates or maybe already had them. There was a lot of preliminary work to do and I thought he was just now starting the last step of the job, which was to gather a list for his select group of electrical engineers.

Leaving then with the call back people was my safest choice. I didn't think Von had seen me, and I might need to appear on the scene of one of his sessions at another time. If I got away with my invasion into his business today, I might still be able to do it again on another occasion. Mixing in with the others I left the auditorium and went back to my car.

I turned on the engine, opened the front windows, and turned on the air conditioner. After the air cooled and got circulating, I closed the windows to give myself the privacy to sit back and think about it all. This was the way he was getting his customers, so he didn't really care about getting them through inquiries at Vansar Electronics. He was going to supply electrical engineers with a way to get into molecular manufacturing using, his own Vansar Computers. The logic of it seemed to be that they would be using the Vansars to learn how to set up a molecular manufacturing machine. I didn't know if he was going to sell them stock in Vansar, or what the actual business arrangement would be to put them under the right legal umbrella. I did think he would sell them shares of the company to go along with his consultations. And the thing I hadn't counted on was that his roll in it was to be a legal consultant. I thought whether he was a shark or not, he was smart enough to be a legal one. It all went together and he would probably pull it off. He certainly must have elicited the services of a lawyer to set up the paperwork for his select group of engineers. He had to be sure everything was legal for them, and for himself before he had them sign the bottom line and write him a check.

There was one big thing about it that I couldn't swallow. I wondered how Von could really deliver the goods to these engineers, as I had been with the Comp-ease group for quite some time now and they hadn't been able start up a successful molecular machine in all this time. I didn't think Von could do it either. He was just trying to get something back for all the time and money he had invested in his very special education. He must have thought he had a right to sell his skills like a carpenter who builds a house. The carpenter who became a contractor could sell his completed house for a good price, but it wasn't that simple for a nanotechnologist.

Calling the Comp-ease factory number on my cell phone I caught James Lebeau still in his office to give him my report.

"It looks like I've opened up a can of worms, James," I said. "The man behind the Vansar Computers scheme is employed in one of the top positions at Gladstone Electronics. It's possible my job could be terminated if I pursue it."

"Well, I told you I'd pay your fee," he said. "I'm not surprised at the trouble being at Gladstone."

"It's not Harold Remkin."

"I didn't think it was, but I've heard he and Von Mueller weren't getting along."

"You heard right," I said. "I don't think Remkin's going to fire me, but you never know. I just wanted you to know where things were leading me."

"Sure, it's leading you to Von Mueller," Lebeau said. So what can I do to help?"

"I'd need to have someone under cover to attend a presentation like the one I just observed today. He's smart and knows what he's doing, but when one of his understudies finds out the table top molecular machine doesn't work Von's going to be in trouble. When they all find out it doesn't work, he could be charged with fraud."

"And he'll deserve it," Lebeau said. "That's why we know to work to see that people like Von Mueller don't hurt the rest of us. We're all on shaky ground right now."

"Could you call Captain Torres and tell him the situation, see if he can have a plain clothes officer enroll into Von's group to document what's going on. I can meet them at a motel coffee shop on the north drag, at the corner of Guadalupe and Jones Drive."

"I'll call him right now and then call you back."

"I'll be right where I am in Austin waiting, my cell's on."

In a few minutes I got the call back from, Lebeau. He said Torres was coming with his plain-clothes-officer to be sure it got started right. I called Joe Torres and said I wanted to fill them in. "I know I've been calling on you a lot lately Joe, I just wanted to level with you," I said. "I could have tried to bring in the Austin Police Department, but it would have taken time, and this is really a part of the case we started back in your jurisdiction, and if you can vouch for me I can bring in APD also."

"OK Dan," Torres said. "It's a political thing. I don't want to get rid of it. I'd have to work with James Lebeau even if you weren't on the case. The big oil people in Luling have been investing heavily into companies like Comp-ease. We have to protect them or the economy could go down, leaving everybody in bad shape."

"It used to be how the big auto companies go, so goes the nation," I said. "Now we seem to worry more about the big electronics companies."

Lawmen have to always consider politics and local situations, when they decide on a course of action. If we didn't we could be out of a job. It put us on the same page and we were ready to move on. We closed our cells.

CHAPTER 44

While I waited for my troops from Luling to get to Austin, I called Claire to let her know I was in town. She answered right away and I guess I was surprised at the abrupt answer. When I said I was in town, I didn't get the usual response and I knew something was wrong.

"Oh, it's you Dan," she said.

"Yeah, me, what's wrong, Claire?"

"You tell me what's wrong."

"I don't know, that's why I asked you."

"One of our lawyers was on a case in Crystalville," she said. "He stayed overnight at the Crystal Palace Hotel and had a nightcap in the bar. He overheard the owner or manager talking to the bartender. They were talking about you, and the woman that was with you in your room."

"It was the owner he heard talking to the bartender, the owner had been drinking a few beers at my room. He brought the beer over there. He also called the woman who was a close friend and invited her to join us without telling me. The woman left my room when he left, she did not stay the night."

"My lawyer friend had met the woman at Gladstone a few years ago, and I had too. It was Marsha Brooks and I can understand how those things happen. It's just that you never even mentioned working with her at Gladstone. I don't think we should see each other for a while. After all, that's the company you work for, and you're going back and forth around there till the end of the case. Just give things a chance to work out."

"Give what a chance to work out?" I asked. "I took her out to lunch one time and that's it."

"You know what I mean," Claire said. "We should just see other people for a while and be sure about things."

"I'm sure about one thing, and that's about your friend," I said. "He's a trouble maker who wants to make the most out of nothing and make a move on you at the same time."

"Maybe so, but just do as I ask I've got your number and I'll call you when I feel better about it."

"Yeah, I feel better about it already," I said. "I'd come over there right now if I wasn't in the middle of an investigation. I'm meeting a couple of men coming in from Luling in a few minutes."

"Be careful Dan. You can sleep on the couch if you stay over. Just give me the night to sleep on it, please. Would you do that?"

"Sure, we'll talk tomorrow."

It helped me to get through all of that, to know that I had been completely truthful with her. I had to be satisfied with the outcome. Turning around on the north drag I drove back south a few minutes till I came to the coffee shop, pulled over and parked. My friends from Luling hadn't arrived yet, so I went inside and ordered a cup of coffee with one of those big fattening sweet rolls. I got the energy pick-up I needed and felt ready for anything when they came in and sat down at my table.

"Don't mean to rush things," I said. "But we need to go to Janco Computers as soon as we can. Joe, my plan is for your man to claim he's an engineer at Adco Computers." I gave them a three by five index card with the name Adco Computers and the address he would need to fill out the contact card Von would give him. They followed me in the unmarked car they were driving to Janco Electronics. When we got there I separated from them, as I wanted to be near the parking lot ready to follow Von at any time he might leave. Joe Torres was going to keep me informed as to how his man was doing.

I found a spot between the parking lot and the place where the short driveway came out of the Janco property to go back onto the drag. I parked on a side street and waited. Von had been making the most of his time with the group of engineers, who wanted to do all the business today. It wasn't any surprise to me that they were still working in the auditorium, when I returned with my help from Luling. In only a few minutes he came along driving a cannery-yellow-sports-car. The big showdown with the special interests had left everybody a winner. We still had every kind of vehicle and most of them running non-petrol. The greenhouse gasses were under control, we were hopeful that we wouldn't have to get aboard spaceships to get a breath of fresh air on another planet.

Von took LH-1 and headed for Crystalville. I followed along keeping three cars behind him. Wanting to learn more about his activities, I couldn't afford to loose him and at the same time he had to be kept in the dark about what I was doing. I'd thought he might go to Luling to stock up on his Vansar Computers. He had three of the small laptops in the auditorium for his prospects to use on the programmed learning part of his indoctrination. If he was going to get serious about molecular manufacturing, his plan called for them to each have their own nanotech computer.

The freeway lights switched on as the darkness encompassed us. I put on my headlights and could see the light house and Lake Crystal in front of me. The flashy yellow car pulled off to the access road and I slowed down to let him take the lead. I could just follow his head lights. He drove from the access road onto the private loop to the Crystal Palace Hotel, and around to the front entrance. I drove slowly to the right side of the loop, which went completely around the lake. I could see the back of the hotel and the lighthouse. The headlights continued around the left side of the private loop. Von drove on the gravel road I had seen the maintenance truck use to access the lake property. He stopped at the lighthouse, got out and opened the full sized trunk. I pulled my car off to the right and stopped as people often did to enjoy the full view of the lake. I could see him walk a few feet to the door, produce a key and go inside the lighthouse. A light came on inside that lit up the stairway to the top. The rounded plastic windows shined beams of light completely around it like an authentic lighthouse, and Albert had said he left it on all night when business was poor.

It seemed as if Von had talked Albert into investing in his new computer company, and I wondered if that was all he had talked him into. Von came back through the door and carried several of the small silver-colored laptops to his trunk and went back inside to get more of them. I left my headlights on to blind him if he looked my way, but he didn't. He was unconcerned as I drove around to the other side of the lake, parked and turned off my head lights. I thought he would drive back to Austin and still have plans to do more business. My plan was to wait until he was gone and then talk to Albert.

He didn't even bother to turn off the engine and cut off the lights. When he finished loading the trunk, he closed the lid and jumped back in the car. I watched his car lights shining from the road, as he drove back on the access road to LH-1, and crossed over to the south lanes to go back to Austin. Having no concern for Albert, he had left the light in the lighthouse burning and I doubted that he even locked the door. I turned on the ignition and the head lights of my own car and drove the rest of the way around Lake Crystal to the front entrance of the hotel. I parked in a space by the office and went inside to find Albert in his office alone.

"Do you have anything valuable in your lighthouse storage space?" I asked.

"Yes, I would say so, why?"

"Well, I noticed the light is on inside, and you seem to be filled up here."

"You're right, Dan," he said. "I haven't been in there to leave the light on. I'd better check it out, have a seat and I'll be right back." He looked out the window saw it on and went out the side door, which opened up to the gravel road leading to the lighthouse.

Waiting till he walked about thirty yards to the structure, I fell in behind him as he opened the door and went inside. I noticed the door had not been locked.

"Have you been robbed?" I asked.

"I guess I have," he said. I noticed he was visibly shaken and seized on it.

"You're involved in more than you can handle, Albert," I said. "Von Mueller has been using you."

"You mean I . . ."

He seemed to be in shock, blurting out things without knowing what he was saying.

Two empty steamer trunks with open lids were visible as soon as you stepped inside. I thought that a lot of those Vansar super-laptops could have been fitted into them. I closed one of the lids and helped Albert over to it so he could sit down on it like a high stool. I had broken him down, not thinking about how the loss of someone else's expensive computers might have jarred him. Now I had to build him back up again.

"Just sit there and try to relax some," I said. "It's alright, they haven't been stolen. Von came in here and got them, and drove back to Austin with them."

"But you I . . ." He was still shaking and stammering.

"Calm down and sit there a while, then we can go back to your office and talk."

"How you know Von came and got them?" He asked.

"I followed him here from Austin," I said "Now you ready to go back to the office?"

"Yes, I'm ready." He said and pushed himself up from the trunk and I supported him as he flipped the light off. He leaned against the door frame and he grasped the lock lever from the rear of the door and locked it, slamming the door as we went outside.

A cloud of darkness seemed to follow us, as we returned to the comfortable, well lit room. Albert sat down at his desk, opened the bottom drawer on the left and removed a half full pint of scotch. He offered it to me and I took it and walked to the cooler where he kept the sodas. His eyes followed the bottle as I picked up a ginger ale and a glass.

"Did you know eighty per cent of the people in prison had an alcohol problem, before they were tried in court and given a prison sentence?"

"No, but I need a drink, Dan."

"I'm going to give you one, but I'd like to slow it down, some. Can we add soda?"

"I can take it that way," he said. "Will you have one too?"

"No thanks, I'm not having one," I said. "I'd like you to think about that statistic again, it's a very large amount, eighty per cent of all those in prison."

"I get the message, and I don't doubt it's eighty per cent."

Pouring about two fingers of scotch in the glass, I mixed it with a liberal amount of soda and handed it to him. He drank about half of it and sat the glass down on his desk. Then he took a deep breath, gathering himself together. "What you're saying is that I'm at the place where I could get myself straight, before something disastrous happened to me."

"One thing for sure about you," I said. "You're no dummy, Albert. You already have your own situation well analyzed, so are you going to get straight?

"I'd like to get straight, but I don't know what to do. You can see how Von treats me. I don't know how you know about it all but you're right. He's a madman."

"It's my job to know what he's doing," I said. "I followed him here from Austin, and if nothing else he has a huge conflict of interest, since he works for Gladstone and is selling shares in his own company, Vansar Computers."

"I'm going back to Austin, now," I said.

"That's where he lives, so better watch your back too."

"Do you have the address and phone number?"

Albert gave me one of Von's business cards and it was written on the back. Taking a big swig from the bottle of soda, I took it with me to my car to finish on the road. Some people had to have a mixed drink on the road. That was not only a bad habit to get into from the standpoint of becoming an alcoholic. It wasn't good old boy country around here anymore. The police might give you a ticket and you could spend the night in jail for having an opened bottle of anything alcoholic in the car.

Just as I started the engine and adjusted my window to allow a little fresh air in at the top, I could see a figure moving toward the car in my periphery vision. It was Albert again and I let the window down the rest of the way.

"Have you thought of getting a permit to carry?" I asked.

"Do you think they'd let me have one?"

"Yes, see Charlie, it's a simple form to fill out. Tell him I suggested it."

This time I really went for it, kicked down the accelerator, made the turn around to the south lanes and was on my way to Austin. It was an easy trip. I didn't intend to use that room at the Crystal Palace, ever again. No wonder Claire was suspicious about it.

Thinking about how things develop and keep changing, proved to be a complicated topic I could use for many trips. Who'd have thought that the indelicate few moments which elapsed from the time Albert went back to the bar, Marsha went to her car, and I fell into bed could have been a major incident, reported to Claire. It had all happened in about a minute. My first reaction had been correct. I thought Crystalville was in the greater Austin area and I could commute back and forth every day.

CHAPTER 45

It was after midnight, when I came to the anomalous fieldstone apartment building. The exterior walls were made with large, flat-cut-stones of different colors, set into concrete instead of plaster like a fresco. Driving up to it brought your emotions into play, and made you feel the same as you might feel about a work of art in a museum. I'm not an artist but I had that idea every time I came there, so it must have been doing something for me.

Letting myself in with the key and finding it quiet, I shed my shoes and tossed the blanket over myself as I wiggled around on the couch. I was tired enough to arrive at the comfortable spot in only a few minutes before sleep settled into me. I slept so soundly, that it seemed like I had only closed my eyes, when the morning sun entered the room and I could hear Claire in the kitchen. I quietly folded up the extra blanket and put it back in the cupboard where it was kept. This was a delicate time for us and I was careful to go slowly, and think about everything that came out of my big mouth.

"The coffee's made, and I'm finished in the shower," she said. "You do your thing. I'll see you back here. "She went in the bedroom and closed the door. I knew better than to rush things.

While Claire got dressed for work, I took a quick shower, poured my coffee and took a hot roll from the basket on the table. I took my light breakfast back to the parlor and sat down on the couch. When she came back in the room with her coffee and roll she smiled as if nothing had happened, but I knew that it had. I had already eaten my roll and on my second cup of coffee. It was clear that she expected me to do most of the talking, it was all about me.

"Well, I don't blame you for feeling the way you do," I said. "She's divorced and I think she did have some designs on me. Something could have happened, if I had let it, but I didn't."

"Oh, that was exactly what I wanted to hear, Dan, but go ahead."

"She has an MBA and was into everything at Gladstone, so I had to be cordial, especially at the beginning of the case."

"Now we're getting someplace," she said. "Cordial is the key word here, when you're like that the poor girl had to love you."

"See, I meant to be a friend and I guess I did help her get through this hard time in her life, but the price has been too high for me. I should have turned down the free room."

"No, you shouldn't have, Dan, you could hardly hold your head up. I know you lost several pounds."

Never having been one to claim they understood women, at least, I knew honesty would help, and it did. My woman was so smart, that all you had to do was give her a little start into anything and she would see right through it. Lucky for me I had the truth on my side. And I had her back on my side again.

"I'm glad you feel that way about it, but I already decided to just commute back and forth to Austin from now on."

"Yes, that's up to you, Dan," she said. "I understand it now and you're forgiven."

It really felt great to be forgiven and it must've felt good to the forgiver, too. We went into a clinch, got stuck together and ended up in the bedroom for a while before going to work. She had to freshen up again to look professional on the job. I understood that. Not so, with me though, and I liked to have the smell of her on me, so I just put my clothes back on and took my coffee out the door with me.

I went to the car and found Von's business card that Albert had given me. Looking at the address on the back of his card, I saw it was in Hyker Park, where Wallace Norton lived. I wanted to check it out so I took Caesar Chavez to Red River Drive, turned left and drove north to the campus. I wondered if they had both lived there as PhD candidates, and had liked it so well that they just stayed on after they were hired for their good jobs. It was still another surprise to find they lived on the same street. The big difference between them was that Norton was married and had a family. Von Mueller lived in one of the small, one bedroom cottages nearer to 'the drag'. There was a dreamy thing between the campus and Hyker Park.

Through no fault of his own Norton had been an early suspect in this case, but now that I could see a connection between Norton and Von Mueller I was starting to forgive myself for my early mistakes. I never trusted either one of them and now they were living on the same street. There was some connection even though there was a big chance I'd never find out what it was, or be able to relate it to this case.

The flashy yellow sports car was missing from Von's small white-cottage, so I had no choice but to call on Norton. I'd catch up to Von later. It was Norton I wanted to talk to first. I drove on down to the other end of the street and saw Norton coming out of his house. I parked at the curb, opened my window and yelled through it.

"I was looking for Von Mueller, and I've missed him," I said. "You must've noticed he's spending a lot of time, doing speaking engagements around Austin."

"Oh, not really, we speak sometimes when we're going out in the morning, but I thought he was still at Gladstone, and driving up there every morning."

"I don't think he'll be there long, since he's got his own computer company. I'm trying to catch him to see if I can invest in it, while I'm making good money. I'm getting a bonus and I'm thinking of putting it all into Vansar Computers."

"You sound like you're serious about that," he said. "I walk around the neighborhood every morning before my first class if you want to go along."

I said I would and joined him as he turned to the left on the sidewalk, by my car. We walked on past the front, corner-lot of his grey-brick house, and turned left again at the corner. It seemed like we were going for a walk around the block.

"Don't you think it would be a good investment with Von being one of the experts in nanotechnology?" I asked.

"I don't really know about that, never heard of Vansar Computers."

"I've attended one of his speaking engagements here in Austin, and he's got a lot of electrical engineers interested in it," I said. "I thought you would have heard of it and maybe even invested in it."

"Look, we live on the same street and we're business associates, but we're not friends. We used to date the same girl, but I'm the one that got her and married her. He wasn't a sincere person, and it seemed as if he wanted her as a conquest, not to marry her. She said he had never talked to her about marriage, but when we got married he took it as a personal affront and didn't even speak to me for a long time."

"So you still barely speak, and never have any friendly chats," I said. "You just speak if you come upon each other and talk about the weather."

"Yes, right, I know nothing about his personal business."

It turned out I was right about his morning constitutional, being a brisk walk around the block. When we completed the block at the front of his house, he went inside for a while before going on to his first class. I climbed back into my car and turned on the engine and the air conditioning. Sitting there a few minutes to think proved to be a good idea. Norton came back outside and asked me if I had time to give him a ride to his class so his wife could keep the car to go to the store.

"It's only about a ten minutes drive from here, but she wasn't ready," he said. "I didn't want to be late. It's a small class and I'm afraid they might walk on me."

"No problem," I said. "You helped me and I help you back."

"It was all true, what I told you about Von Mueller."

He directed me well from one traffic light to another, and when to be ready to make a turn ahead of time. It was when he mentioned Von Mueller again that it came to me. If my wife had dated Von Mueller before I married her, I wouldn't trust him to live so close. I thought to put another question to him before he got out of the car.

"How long have you been married?" I asked.

"Three years."

"If you made it that long with him living so close, it's going to last." I said, not believing it. He had a pained look on his face as he got out of the car. I didn't think he believed it either.

It seemed likely that if something was going on between the two of them, the shopping days would be when it happened. Perhaps her timing was off this morning, due to her excitement about the meeting. I wasted no time getting back to that neighborhood, and when I drove by a block away from the house I could see the blue BMW was still there. I backed up into a vacant driveway and waited. She came out in a few moments and hopped into the freshly polished blue car. I followed it, keeping at least three cars behind.

She drove to Guadalupe, took it to MLK and from there to Lamar Street. By then I knew the way and, when we turned north onto LH-1, I thought she didn't need to go this far to find plenty of good stores to shop in. We were on our way to Crystalville, and yes, I was surprised and saddened about it, when she pulled off at the exit to Lake Crystal Drive and the private loop that ended at Gladstone Electronics. She drove into the rear of the parking lot and parked in a spot that was the farthest point in the lot away from the Gladstone building. I was parked at the side of the loop that was called Gladstone Drive about seventy yards away from her car. I took out my small field glasses out of the glove compartment and made a preliminary adjustment in the lenses for that distance and sat them down on the top of the instrument panel to be ready.

The spiffy, yellow sports-car came along in a few minutes and parked in a space next to the shiny blue car. He got out and opened his trunk at about the same time that she got out and opened her trunk. I picked up the glasses and adjusted them to be able to see what he was carrying over to her trunk. He had three small laptops stacked on top of each other like he had carried the small Vansar Computers from his speaking engagement at Janco Electronics. As he arranged them in the trunk I was able to see the Gladstone logo on one of them. He went back to his trunk and she followed him. They each carried an armload of the Gladstone Computers, fitted them all in the trunk of the BMW, and then went back to their own cars without a word. He drove his car to the front of the lot, parked it in a spot reserved for the administrators, close to the building, and he went inside.

I was sitting there, stunned, in my own car while she drove her car to the access road and turned around to the south lanes of LH-1. I thought I had just witnessed computer theft of such magnitude, that it could only have been accomplished by those in management. The only thing I could do was follow her and try to see what she did with the nine super-laptops. It all started with one and now there were at least nine.

Tromping down on the accelerator pedal, I caught up with her and maintained a safe distance the rest of the way to Austin. When she got off the freeway at Austin she crossed over and drove south on Lamar Street, turned left at MLK and left again onto Guadalupe. I followed her going north away from the university to the North Drag high-tech compound, where she stopped and parked in the Janco auditorium parking lot. Not waiting for Von, she made three trips to take the nine computers inside.

CHAPTER 46

My car was legally parked on the side street and I got out and locked it. It suited me better to be able to make quick checks in the auditorium without moving my car around, and if I wanted to follow somebody it would be easier. In some situations it was better to be on foot and this was one of them. A small crowd was starting to form around the auditorium doors. Something was going to happen, but I couldn't see how Von could do another nanotechnology speech, at the same time Janet was working with nine people on computers doing programmed learning.

Cutting through the parking lot to get to the auditorium, I stopped as Joe Torres got out of his unmarked police car. We met in about the middle of the lot.

"I just observed our subject, John Von Mueller remove nine Gladstone Computers from Gladstone, to the car of his woman accomplice. She took them in the auditorium to use with a group while Von worked with another group"

Von Mueller drove into the parking lot into a spot closer to the auditorium than where we were, and parked in a space someone had just left, got out and went inside. I continued filling in Joe.

"Von's inside getting ready to start his speech," I said. "I believe we have enough on him now to make an arrest. There are two of them in there, and with the crowd it might be hard to get them, with just the two of us. Since it's going on in Austin now, do you want to bring in the APD?"

"Yes, I'll handle it," Joe said. "I'm sure they'll give us a couple of officers so we can make a safe arrest, and use their jail while we write up all the formal charges."

"I also have a witness in Crystalville that will prove Von Mueller's involvement." "I don't need to go to the APD office," Joe said. "I know people already on this case, since the deer rifle killings."

While Joe made the cell call I went inside the auditorium, noticed that tables and chairs had been placed along the left wall and seven people were using computers. Janet was moving around between them occasionally answering questions. Von was standing at the front testing the microphone, almost ready to start his presentation. I was wrong in my assertion that both things couldn't be done at the same time in there.

He was talking about some of the longhorn athletic programs like before, and I thought that's where I came in, so I turned around and went back outside to conclude the plans for their arrest with Captain Joe Torres.

Joe had made the arrangements with the APD, and three of Austin's best plain clothes officers were on their way over here. The man he brought with him from Luling was inside doing the programmed learning in Janet's group. His experience in electronics was as a technician, but he had come in late in the last session, filled out the contact card and had passed as an electrical engineer. We would copy the serial numbers of the computers in front of him and let him keep a copy of them so he could be a witness in court. We'd put a circle around the serial number of the one he'd used, and do the same for the other six people that had used those computers.

"We'll have to check the numbers at Gladstone tomorrow," Joe said.

"There's no way he just borrowed them," I said.

We were prepared to present a strong case against John Von Mueller even without the Gladstone Computer evidence. I felt confident when the APD officers drove into the parking lot. We met, identified ourselves, and discussed the evidence to be protected in the auditorium. One of the Austin detectives was a Captain and the other two were Sergeants.

When the five of us went inside together, we might as well have worn signs that said cop on them. Weighing in at one hundred ninety pounds, I was the smallest. Von had finished his speech and most of the crowd was gone. Joe and I went over to Janet Norton, informed her of our plan to make arrests and asked her to sit down at one of the tables. The man who was our plant started working with Joe to write down the serial numbers of the laptop computers. I started the paperwork on Janet and took it with us when we walked down the center aisle to arrest Von.

He looked straight at me, his hands together to be cuffed as the officers circled him.

"You'll never pull it off, Brandon," he said.

"I've already pulled it off, Von," I said.

They took Von out to the patrol car and locked him in the back. The driver waited with him while we went back inside to finish up the paperwork on them, and bring the computer evidence. I told Janet Norton to cooperate and she might get off.

"I had no idea he had stolen the computers," she said.

"He and the CEO weren't getting on well, no way he'd borrow them."

The APD provided tags to put on the computers and forms to give the contact information for the two suspects. We waited till we had it all ready, a tag on each computer with it's serial number written on it, then we put it all in the patrol car and went to the downtown booking office with our suspects.

I made my lengthy, written statement, at the APD, and I let them give me a photo-copy of it to help me remember all the things I still had to do to support it. This was getting to be a long day, and I had to go to

Crystalville and get a statement from Albert Hoffman to document what he had told me about how Von Mueller had orchestrated everything regarding the theft of the super-laptop and the deer-hunter shootings. I typed all of that information about the relationship between Albert and Von Mueller, especially everything he knew about Von Mueller. They let me use a police computer, so I could get duplicate copies if we needed them. I printed three copies to take with me to Crystalville. I left plenty of room for Albert to sign it.

Joe Torres drove me back to the North Drag to get my car that was still parked on a side street near the Janco auditorium. We talked about his man he had planted.

"Did he have any observations about the programmed material?" I asked.

"Yes, he said it was like being brain-washed."

"Can we produce a copy of the materials?"

"Sure we can," he said. "It's all saved in the Gladstone Computers just like it was saved on the Vansar Computers, so we could have it all printed from one of the computers in evidence at APD headquarters.

"As soon as you drop me off at my car, I'm going to Crystalville. Would you give me a copy of the computer's serial numbers?" He smiled and handed me a print out of the nine serial numbers.

"I bet you're going to need to get back to Luling, Joe," I said.

"As soon as I drop you off, I'm on my way home," he said. "But remember I'm just a cell-call away."

When we came to my car, we shook hands over our success, and I jumped out of his car and into my own. I drove south on Guadalupe to MLK and west on it to Lamar and the crossover to LH-1. I was glad to be able to tromp down on the accelerator without having guilt feelings like in the old days.

Going first to Gladstone, I traveled to the exit and drove directly onto the private loop to the parking lot. I parked and went inside to the new reception area, thinking I might be in the wrong place, since several other companies had the same identical reception layout. It was a long day but

I was just where I wanted to be. It seemed as if they were still improving the security system, because I heard the click of the relay switch come from under the counter and the gate automatically went up. The voice of Carl Watkins said, “Come on back Dan.” I went through it and it closed automatically behind me. “How’d you like that action?” Carl said.

I continued across the corridor and I could see Carl’s head sticking out of his office. I waved and said, “It worked fine.” He ducked back into his office and I followed him in. He motioned to the coffee and I served myself a cup and sat down to the side of his desk. “They’ve got the same set up in the offices upstairs,” he said. “Say, if Remkin sees you on his monitor in his office, he can do just as I did, or come to meet you and let you in manually.”

“You’re getting more choices around here every day,” I said.

“So what can we do for you?”

“Can you trace Gladstone Computer serial numbers?”

“I don’t have to move from my desk.”

Handing him the print out of the numbers, I leaned back in my chair and took a big swig of the coffee. He had a large desk with a laptop computer and all the peripherals. There was a strip console with buttons to the right side of the desk. He put the numbers in the system, punched around a while and the monitor of the laptop filled with numbers.

“This series of serial numbers isn’t among the numbers of those laptops that have been sold recently. Hang on I can fax them to Marsha. She’s got warehouse and all.” He went back to the computer and sent the numbers over the fax to the office on the third floor. Then he refilled his cup with coffee, brought it back to his desk and got comfortable. “What’s this all about?” he asked.

“It’s about nine of those expensive Gladstone super-laptops reposing in the Austin Police Department’s evidence room.”

He pushed away his coffee cup and sat up straight in his chair. “They’re stolen computers?” he asked.

“Unless Von was allowed to borrow them, or they were a gift to him.” The phone rang and Carl answered it. He listened for a few minutes, said thanks and hung up.

"Those computers are supposed to have been stored right on this floor at the rear of the building," he said. "Come on we'll go over there."

Going as far as we could to the rear of the building, we came to a locked door to the right. Carl produced a key ring from a clip on his buckle, and unlocked the door. He flipped on a light and we went inside. I noticed there was also an ordinary sized door at the rear of this room. "Does this door have a driveway on the other side?" I asked.

"Yes, it does," he said. "This other locked door to the right goes to the assembly line, and we've had a video camera over there for a long time."

Carl led me to several tables that were backed up to about three feet from the outer wall. They all had computers lined up on them. I walked back down that aisle to the door to the outside and turned left to look at that table. There were outlines of dust for the nine computers that had been taken. "Here's where they were," I said.

"You knew to look there because you thought someone drove up next to the door and opened it with a key," Carl said.

"I'm sure that's what happened, because I saw Von transfer nine computers from his car to his girlfriend's car. Then I followed her to Janco Computers, where she took them out of her car and put them in the auditorium, where they were confiscated by the APD earlier today."

"Wow, you been busy, Brandon," he said. "I'm going to have the supervisor inventory all the computers back here so we can prove in court that they've been stolen."

"I think you've got it."

"A lot more than you think," he said. "If Von did it within the last two days we've got it on video." He pointed to the wall that was shared with the assembly line room and I saw the video camera high up extending from that wall.

"It was put in two days ago so he probably didn't see it," I said. "There he was making a mockery of the company still paying him, while he ran around doing his own business on their time. It's ironical that it was what really caught him."

Carl didn't take long to check the video camera and we saw Von perform the theft alone. He came in the door quickly and left the door open while he made three trips to the car and had the presence of mind to lock the door when he left. Von knew exactly what he was doing, but not that he was being watched. Carl made a copy to send to the APD and I thought there was no hurry. If it got lost Carl would just make another copy. I didn't feel like we were in a crunch now, but Harold Remkin had gone home before I arrived. We walked back to Carl's office to reclaim our chairs for a while.

I found Remkin's home phone number on his card in my wallet and punched in the numbers on my cell. He answered in a startled voice that made me think he had been taking a nap, and that I would need to go slowly.

"I'm sorry to bother you at home, sir," I said. "It won't take but about a minute for you to clear something up for me and I have something to report to you."

"Oh, it's Brandon, sure, go ahead."

"Is there any way John Von Mueller could have made arrangements to take nine of the super-laptop computers from the assembly line storage room to be used in his own side-line business?"

"No way, he'd have to get it in writing from me to go to the stockholders," Remkin said. "If he took computers out the back door he stole them, and should go to jail."

"Well, that's what happened and he is in jail," I said. "I'm sure he managed to get keys to everything at Gladstone. He got a key to the computer lab and took the original super-laptop. From then on he was involved in grand larceny and murder."

"Thanks, for calling me to give me the good news, I mean it. Von needed jail"

"Do you agree then, that this case is solved," I said.

"Yes, I certainly do, Brandon," he said. "With Von Mueller gone I'm going to see if we can get Bob Clauson to come back with a promotion into Von's vacant spot. He wasn't doing anything but larceny around here anyway. I'm sure Bob would do better and we'd also have his expertise."

"You'd need to talk to Bob about that, but I know he has a bad case of agoraphobia and he stays in his house most of the time, and they go to him, and he has Steve Wingate do most of the leg work for him."

"We'd put him on convalescent leave, and he could do the same here," he said.

"I'm sure Nelda would like them to work in the same town, I still have to go back to Luling to wind up some things, and I'll pass on your offer if you like."

"Yes, I'd like that."

"I'll get together my expenses and record of activities to submit to you tomorrow."

"That'll be fine," he said. "I'm going in at eight o'clock."

"There is just one other thing," I said. "Are you still the only one who knows the combination to the safe in your office where the research lab keys are kept?"

"No, Carl Watkins knows it and he's probably there now, if you had in mind taking the extra key to see if it matches a key Von Mueller might have."

"Exactly what I wanted to do," I said. "I'll return the key tomorrow."

It was all fine with Harold Remkin and we closed our cells. He had been overbearing at times on this case but for the most part he was fair and honest and what more could you expect from an employer. Carl was quite intent on my conversation and I wanted him to hear it to cut down on the time it would take to tell him everything.

"Remkin said I could take that extra key, so I could match it to one we might find at Von's in Hyker Park."

Carl had access to everything that had anything to do with security. We went up to Remkin's office on the third floor and he worked the combination to the lock on the vault and opened it. There was still a key inside along with the petty cash. He handed the key to me. Carl liked the sound of my plan and especially the part about Von Mueller going down hard. He heard me say I'd be coming back to make final preparations at Gladstone. We said we'd see each other then, but cops never want to say goodbye.

CHAPTER 47

I went out to my car and drove about a hundred yards around the private loop to the front of the Crystal Palace Hotel. I took a folder along with the summary of what Albert had said about how Von had set him up from the beginning of their relationship. I found Albert in his office behind the registration section and went in.

"Have a drink or whatever and come sit down," Albert said. "I have something for you, Dan." He handed me several sheets of hand written documents with his signature on each page. "I had our lawyer check it and he said it'd stand up in court," he added.

When I compared it with the summary I had already prepared, I wadded up my summary and threw it in the trash can by Albert's desk. The statement he had prepared was in more detail, with dates and time of day added. He had stated the lies that Von had told him. And how he had been threatened and set up to do things all along by Von. He also included statements of others that were contrary to things Von had told him.

"This is much better than I had hoped for," I said. "Did you copy it?"

"Yes. My lawyer has a copy of it too."

Albert even had documented how Von had bragged about turning Bob's three friends against him and got them to beat him up.

I still have to take this to the APD, so I hope you'll forgive me for rushing off now, but I've got to get back to Austin."

We agreed that it would be better to get together to visit on another day, when we had all of this behind us and could enjoy it more. I didn't know when that would be though.

I went back out to my car, drove to the turn around and to the south lanes of LH-1. It had been dark for a while already and I still had much more to do than I would have liked. I kicked the accelerator down to the floor and left it there until I saw approaching signs of Austin weirdnesss. I dropped down the speed to avoid a ticket. It would be too much time lost to wait for a patrolman to write me one so I cruised at that speed the rest of the trip. Pulling off LH-1 to the crossover to Lamar Street I took my usual route south to MLK and from there to Guadalupe (the drag) and to Von's small rented cottage in Hyker Park.

Driving up the driveway to the rear of the house, where there were low-hanging trees to afford more cover, I got out quickly and went to work on an outside window with my Swiss camping knife. I shoved the blade in the open space at the top of the bottom pane and pulled it into the locking lever in one motion. I was standing on an outside porch, so it was easy to open the window, sit on the window ledge, and shove my legs through it to get inside.

As I went through the house I looked at table tops and went through drawers. I went in the small bedroom and opened the top drawer of a chest of drawers next to the bed and there it was. The key ring only had a few keys on it. There was only one on it that visually compared to the one I had brought from Gladstone. I took it off the ring and put the two keys together I tuned on a lamp on the top of the chest and held the two keys up to the light. They passed that test but I had still a better test that I had learned a long time ago. I had brought a thickness gauge and a six inch machinist rule with me. I could measure the slots in the keys and pick up the slightest differences between the keys. They were identical so I put the key back on the key ring and put the keys back in the drawer. I'd bet that his key was a match.

I went to the window, closed it and moved the lever to lock it. Setting the dead bolts on the door to lock behind me, I went through it and slammed it shut. I was back in my car backing out of the driveway and the venture hadn't taken but about three to five minutes. Driving back on Guadalupe south to the downtown APD building I asked for Captain Williams who had been in on the arrests we had made. It had been such a short while that he still remembered me. When I gave him Von Mueller's address it even rang a bell with him. He didn't question me when I told him I knew the key evidence was in the top drawer of that chest of drawers.

"What you say I have a surveillance team start watching that place right now, till we can get in there and go through it." He made a quick cell call to set it up.

That would be good enough I thought, as I handed him Albert's statement to be included in the evidence. He was a big man with blond, clean cut features, quiet and careful enough to be thorough. "Let's take it to the evidence room," he said. I appreciated his including me, so I went along.

"I remember your statement mentioned the missing key at Gladstone," he said. Do you have another key from Gladstone we can match it with?" We came to the storage room and I noticed everything was labeled properly. I gave him the spare key from the vault in Remkin's office. He knew I had the key and planned on putting it in the evidence room. He added a tag and labeled it, placing it on a table with the other things.

"There's a documents room where I'll put his statement, "he said.

"I noticed he had it certified, but of course, he'll still have to testify." The statement was preponderate evidence.

Assuring him that I was aware of this being the kind of important case that would have lots of witnesses, I said I was also prepared to testify. Von Mueller was more than just a bad guy. He was well-educated, cunning and hard to bring down. He had taunted me, saying I would never pull it off. I remembered what a job it was and I didn't want someone else to have to go through all that for a long time. His education and intelligence couldn't compensate for his being a sociopath. He would be a repeat offender and I hoped he'd be linked to the murders and go down for a long count. In his statement Albert said that Von Mueller had forced him to get the rifle back from the garage so Von Mueller could pass it on to the man Von Mueller had selected to do the murders. Albert had stated in his certified letter of evidence that Von Mueller feared the three young men that were killed, would identify him as the one behind the entire scheme. Albert also stated he attended another meeting where Von Mueller had promised to reward them, when he had sold the first super-laptop he had taken from the Gladstone Lab.

Captain Williams was thinking the same way I was about putting Von away for as long as we could. We talked about it for a while and then decided there was still a lot to do. I left him as he was going into the documents area, and walked back out to my car. I counted fifteen hours of running back and forth today, and I thought it would be enough, so I got back on my iron horse, started the engine and drove to my favorite apartment development only a few blocks away. The colorful outside walls beaconed to me and I found a place to park, not too far from the entrance and went inside to Claire's ground floor apartment. I let myself in with the key she had allowed me to keep, even when I had been under suspicion of being untrue.

Claire was in the front parlor reading something in a large hard-back book, as I came in the door. She dropped the book on the sofa and met me about mid-room. We didn't have to say anything. It just felt good to squeeze each other for a long time.

"You look bushed," she said. "We'll have some enchiladas and eggs." It sounded great to me, in fact, I salivated but silently.

"Sit down and put your feet up on the stool there. Be ready in a few minutes."

I did as she suggested and when she brought it in she put the hot plate down on a TV tray. Then she brought her own TV tray with a hot plate of the same and sat down beside me on the long couch. I thought she must have been waiting for me, as she shoveled the food down as fast as I did.

"Did you know I was waiting for you?" she said.

"I thought maybe, came soon as I could.

"I bet you did, one of those sixteen hour days, huh?"

The last three days I only was able to have a few cat naps in the front seat of my car. "Have you been preparing for your case?" I asked. No need for me to tell about my pro

"Only ten hours," she laughed.

When we finished eating, I jumped up and took the dishes in the kitchen. She said I could just leave them in the sink, but I let the automatic dish-washer do them. It went through a special rinse cycle to remove any excessive food. When it was finished that cycle the water and food were removed. The water was purified so it could be used over. It saved a lot of water and I was impressed, since this case had made me aware of the on-coming water shortage to be caused by global warming.

I returned to the parlor in only a few minutes. At such times, Claire liked to clown.

"Oh, my kitchen hero's back," she said. "No clogged sinks, or such, in my kitchen."

It was time for me to put on my tickling act. I came and sat back down beside her on the couch. She never could stand to be tickled, and I sat there innocently for a few seconds, then I suddenly threw out my hands on either side of her, twisting at the same time. I never did anything but say tickle, tickle, tickle, and she'd jump and pull away. It was one of our harmless little rituals that she always liked. It brought out the little girl in her it seemed. So with that over with, we got down to some serious things.

"I could tell from what you said about other lawmen coming in from out of town, that you were going to be seeing some action, Dan," she said. "I'm glad when you prepare me like that, and it's sure good you didn't get hurt."

"There could have been some action, but it turned out that only two arrests were made and things were so well planned that it was easy."

"So are you near the end of this case?"

"Yes, I think so, but I can hardly believe it because there are still so many loose ends," I said. "Most of them are in Luling, and I've got to go back there tomorrow."

"When this one's over go back to divorce cases. We don't need the money."

"It's not about money," I said. "I just don't like divorce cases."

"And we should like the work we do," she said. "We've been here before, so I'll shut up. Things change and maybe divorce cases will get more interesting sometime, so keep it in mind."

I agreed with her about change and said I'd change my mind, if the time came. She put on the TV news for a while and got her mind off things that probably wouldn't happen anyway, and I was glad to help her. She poured us a glass of sangria and we loosened up and talked a little silly before we retired to the bedroom.

Waking up at about the same time, we decided on the continental breakfast. I think we had talked too much about our jobs and our future. You have to do that sometimes and I promised her I'd take her on a vacation, when I finished this case if she could work it out on her job. Claire liked the idea and thought she could do it, so we had fun talking about where we'd go and what we would do. The day before had been long and a real bummer for us. The getting together to talk about the future, getting over our short spat, and making love were all quite redeeming, as we had been physically and emotionally drained. I think she wanted us both to get out there early, get started on the day, and that way I'd finish my case sooner and we could make our vacation plans. I felt the same way about it.

CHAPTER 48

It did seem that I was wrapping things up on this case. I was driving south on Highway 183, thinking about the questions I needed to answer not just to finish things, but for my own mental well being. Some of the things I had observed about Bob Clauson made me wonder if he was a master of the double speak. It was like he was two or three different persons. I'd think he was one place and he'd pop up someplace else. At first he had been associated with criminals and I never did understand that. The FBI men told me not to worry about it. I had actually considered the possibility of him working for two sides like a double agent. There had to be some reason that he had ended up with the body guards from the mob.

My cell sounded off to waken me from the highway hypnosis which was taking me over. It was Donald Henson, who had taken Nelda's old job at the UT research lab. He was shouting into the phone.

"The police were over at Wallace Norton's house, and here talking to me," he said. "Wallace either slipped in the shower or someone hit him on the head while he was in there. He's in the North Austin Hospital now in a comma. You seemed like you were a friend so I thought you'd want to know."

"Yes, I do want to know about him," I said. "Can you give me the hospital number? Then I can call back myself later and keep up with his condition."

Donald gave me the phone number and I wrote it down. I told him that I appreciated that he remembered to call me about Wallace and we closed the call. It seemed unlikely that Von Mueller had come up with enough money to make bail already. I had slipped in the shower once and I knew you could get a concussion from it. It would be a lot easier to keep up with Wallace by phone, than turn around and head back to Austin.

Arriving back at Luling about nine o'clock in the morning, I kept going straight on the main drag till I came to the industrial center. I pulled over at Comp-ease and parked in a visitor's spot in the front and went inside. Someone saw me on the security monitor as soon as I came in the door and asked me who I wanted to see. I said either Steve Wingate or Bob Clauson, whoever was available. I didn't recognize the voice. In a few minutes Steve came to the gate and opened it. We went to his office and we sat down with a cup of coffee.

"Did you want to see Bob too?"

"Yes, if he's here."

"No, you'll have to go to the Pecan Street house to see him."

When I asked him if he had seen Captain Torres and heard about the arrests we had made, he said that Torres came right over here to tell James Lebeau about it all. That was one thing I didn't have to do now, so I was ready to go to the house and talk to Bob.

"James said that those two were going to cause a lot more trouble for us all if you hadn't caught on to their operation."

"They had been doing it for a while," I said.

"You got suspicious about John Von Mueller, and watched him till he trapped himself."

"Yes, that's what it amounts to."

We finished our coffee and I decided to go on over to the house to check out Bob and tell him about the offer Remkin was prepared to give him. Steve walked with me to the gate, and I walked out to my car. It was a beautiful sunny day with the usual breeze to make it feel good outside. I climbed in started it up and adjusted the window to let in some of the fresh air and drove out the back way to Bob's house on

Pecan Street. When I parked in front and got out Bob was at the door and opened it for me. He stood in front of the open door and shook hands as I crossed through it. I thought it seemed as if the agoraphobia was better.

"Have you been getting out and around a little better?" I asked.

"Yes, I was over at the factory working with Steve, earlier this morning."

Here he was popping around from one place to another with no car in front. I just wanted to say something about it. That didn't gel with me, but for some reason he must have gone over there with Steve for a special job and then Steve had driven him back to the house. "I guess you have plenty of drivers," I said.

"Yes, Sally drove me over to the factory and then back," he said. "She's one of the government people that Steve got for me."

"Isn't she the one that followed me to my surveillance point one day?"

"Yes, it was Sally and Pete," He said. "They asked us all kinds of questions about you, but we assured them that you were alright. It won't happen again."

"It's good to know that," I said. "What I've been wondering, is how you ended up with those mobsters before?"

"That was my doing," he said. "It was the time when those three guys had taken me and beat the crap out of me, and I just picked up Bud Wilson and the others in a pool hall." He laughed and I thought he was telling the truth and I laughed too.

"Junior was bad, but that Tom Oliver was the one to use a sap on me," I said. "They put me out of operation for a few days, so I can talk about it."

"It's good we're over it, so we can laugh." he said.

I had never expected him to be so approachable. Most of the times I had been around him I thought he was an enigma. And it was no different now. He had answered all of my questions and I should have been ecstatic but I wasn't, I was still confused. I pulled myself together and turned around to go back to the door.

"Don't you want to sit down a while, have a drink or something?"

"No thank you, I just had coffee over at the factory," I said. "I don't need to sit down, just one more thing. I talked to Harold Remkin at Gladstone after we had arrested John Von Mueller and he had a pretty good idea, I thought. He said that since Von was gone they could hire you to fill his spot, and you and Nelda would supply all of the technical expertise. He said they only needed one PhD and Nelda would be it. What do you think about it now that Von Mueller is out of the picture?"

Bob seemed to be shocked to hear my message and made no immediate response. "Well I guess I'll have to tell . . ."

"Oh, tell Nelda?" I asked. "She already knows about it."

"Yes, that's good to think about, now that Von Mueller's gone."

"Remkin said you could take your time getting into things, and they would come to you when they needed help like they do at Comp-ease."

"Well, of course we, I mean I'd have to go to talk to Remkin about the details."

"Oh, sure you'd need to do that," I said. "I'm just the messenger and I've delivered the message, so I'm going to get going now."

"Well, thanks, and that sounds like we'll take the deal. James Lebeau is the PhD at Comp-ease and I think they have all the technical expertise they need now. Were you supposed to tell Remkin one way or the other?"

"Not at all, you'll need to talk it over with Nelda and then get in touch with him."

Letting myself out the door, I left the enigma standing there, closed the door and walked out to my car. It seemed to me that I had got off light with my surveillance duties. When the case started to break I followed it north to Crystalville and Austin, thinking the situation in Luling would just fall into place. I didn't even tell anyone here that I was working without a fee, because I didn't want a job now. I had promised to take Claire on a vacation and it would be a real one, not one of those computer transplant vacations that are starting to come into vogue now.

I drove south onto the gravel road and then took the unmarked, primitive crossover road to the place where I had parked before, when I went to the surveillance point. I parked and walked to the wooded area and up the hill trail to the top. It had been a welcome revelation, when Bob Clauson told me he knew about how two of his guards had tailed me over here. At the time I had a lot more to do than finding out whom they were, but now I wouldn't have to do it, and I was thankful for the help.

Taking out my pocket field-glasses, I scanned the house for over an hour. It made me feel more worthy, although there had been no intimation that I had shirked my duties at Bob's safe-house. Harboring guilt, because I was missing something, then I compensated for it by making myself accountable for the safe-house.

It was time for me to stop playing mind games with myself, and I thought that there might be a change in Wallace Norton's condition by now, so I called the hospital. His condition was reported as stable, but that didn't tell me much. I went back to my compensatory activity of scanning the house and grounds. When I focused on the extreme left side where the view was obscured by a wooded area, I could see a bright yellow or orange car parked there. At about that same time I heard a sound that sounded like a shot or a car backfiring. At that distance you couldn't be sure which it was but I thought I should go down there. As I started down the forward slope of the hill, I punched in the number for Wingate's cell. When he answered I asked him if anyone was supposed to be there. He said not, and that he was going to come over there too. We closed our call and I continued on down the hill. As I got to the bottom I could see the shiny car driving away through the trees. It was the first time I had been aware of the visibility of that part of the property. There would be no way for me to get to my car in time to follow it, so I ran to the back door where I thought the person must have come from. The door was open and I went inside to find Bob Clauson on his stomach lying in a pool of blood on the floor. I found a towel there in the kitchen to shove into the entrance wound just below the shoulder in the back. Doing the best I could to stop him from bleeding out, I pressed the towel hard and it helped but didn't stop it. I had only been administering to him a few moments, when Wingate came rushing in. He rushed back outside and I could see him through the back window moving a small white-colored helicopter. Opening the oversized door, I saw him stabilize the aircraft and turn it off. He came back inside and helped me to carry Bob into the two rear seats of the mini-helicopter. A lot of people had them in their garages these days, but I didn't know they had one here. Wingate had the aircraft running now and was seated in the pilot's seat. As we arose above the property, I had to assume Wingate had received the required pilot training. He glanced back at Bob who was still unconscious. "We use this sometimes to get around bad traffic jams," he said. "There was one this morning and I used it to get to work. Holding up alright back there?"

"I guess we are," I said.

"Well hang on we're about to land." It was hard to believe we were there so fast.

At the hospital the nurses and interns took over as soon as we landed in front of the emergency entrance, and had blood, and an IV ready. Moving Bob onto one of the ER beds on wheels his vital signs were checked and he was wheeled on to the operating room table where a doctor pulled out the bullet. Blood types didn't matter now that the use of the proper chemicals had made them all the same. Chemistry had solved most of the blood and water problems. Wingate and I were taken to the room that Bob had been assigned. The surgeon came to the room to talk to us, as the nurses wheeled Bob into the room and transferred him to the more substantial bed in the room. I noticed Bob was still unconscious and when I asked the surgeon about it he said there was no concussion, so he should be coming out of it soon. We were assured that the bullet had not nicked the heart or lung, though it had come close, and a detective from LPD had already received the bullet. As I started to respond to the news about the bullet, Joe Torres came into the room and the surgeon left. Joe provided a guard in the room around the clock 24-7. "I think trouble must follow you, Brandon," Joe said. "What evidence was found is on its way to our forensic lab." I gave him the address of the safe-house and he called it in to the forensic team and they were going to meet us over there.

"Trouble's liked me for a long time," I said. "I think you've met Steve Wingate." They shook hands and nodded to indicate that they did know each other.

"I've also got a friend in the hospital in Austin, Wallace Norton," I added. "He's another PhD in the research electronics business and I know it's related to this shooting."

"It has to do with the arrests we made in Austin," Joe said.

"Bingo, it's the husband of Janet Norton who we arrested with John Von Mueller, PhD, in Austin."

"It's warfare over who gets in first with molecular manufacturing," Steve said. "Those who're afraid our research at Comp-ease and at Gladstone is going to be too much. They're trying to remove the edge we have on them."

"Oh, I get it," Joe said. "They'd hurt both companies by getting rid of Bob."

"Makes sense," I said. By eliminating Bob they reasoned they would be the only ones left that would be able to do Molecular Technology."

"What I've got to do now is investigate the crime scene," Joe said.

"We have the mini-copter, so come on I'll drop off you two cops back there, to do your thing."

Joe told us that every town in the country had their own helicopter, and he happened to know the one from Luling was being used, so he accepted Steve's offer. Steve's car was at the safe-house. He must have driven it over there from their resort-river-house, to get the helicopter and avert the traffic jam. I thought he would probably put away the compact helicopter and take his car back to the hospital to wait for Bob to come out of it. I wondered if Bob had hit his head when he fell or if the shock had caused a psychological reaction to cause the unconsciousness.

The "two cops" got in the seats behind the pilot as we were directed by Steve, the cyborg, and in a few minutes we were hovering over the back yard of the safe-house. Steve landed the small-craft and when we got out he easily rolled it back in the make-shift storage place. He remembered I didn't have a car there so he waited around while we made our preliminary investigation. All I had time to do before was keep Bob alive to make it to the hospital, and secure the crime scene by closing and locking the door as we left.

The large forensic van drove to the spot behind the trees on the left side of the property and parked. Two men and a woman got out. Joe met them at the door and they went inside. I had already gone through it once with Joe, and we found no hand gun. There was the blood evidence they could get the DNA from and that was about it as far as I could see.

Steve drove me over to my car, which was still parked on the crossover street on the other side of the surveillance hill. Then he drove Steve back to the Luling Hospital. Steve asked the nurse again about Bob's condition and she was honest, but evasive.

CHAPTER 49

It was important to know what was going on in Austin and there wasn't anything else I could do to help things in the Luling area. Joe Torres had it under control around there. Steve said he would keep me informed through the cell, when Bob regained consciousness.

Leaving the hospital, I went to my car and drove straight up Highway 183 to the Austin city limits, where it started to take on other names. When it became Ed Bluestein Boulevard I pulled off in a warehouse section where I could park and use my cell phone. I called the hospital number I had been given for Wallace Norton. When the operator answered I asked about his condition. She said that he had been released from the hospital. I shouldn't have been shocked, as it happened all the time. If you could get out of bed and walk you were released. I called the APD office using the number on Captain Williams' business card and he answered on the second ring.

"Brandon, here," I said. "I just got back from Luling, and I need to know about Wallace Norton. He's been released from the hospital."

"You've caught up and surpassed me, Brandon," I thought he was still in the hospital." We checked things out at his house and couldn't find any signs of foul play."

I told him about how the case in Austin was connected to Luling, and how Von Mueller's car had been at the safe-house when Bob was shot. It had to have been Janet.

"It couldn't have been Von Mueller," Williams said. "I know he's in Jail, and we're going to keep him there. His lawyer hasn't even got the judge to set bail yet. It's going to be too large for him to make anyway, so not to worry."

"What about Janet Norton?"

"Had to let her go," She had a great lawyer, got the judge to set ten grand for bail and had it on a money clip in his pocket. She claimed not to know Von lifted those computers. No priors, not even a parking ticket."

"I'm thinking about how far Janet is involved with him, to the detriment of her marriage. I understand there's a war going on between some of the PhDs in competition to be in on the first start up for full scale molecular manufacturing. The company that does it first is going to lead a New Industrial Revolution, which will be the biggest ever."

"I thought that was just a scam," he said.

"It's a scam when they can't do it but claim they can," I said. "When they sell shares in their company and collect money with the understanding that the people are buying into molecular manufacturing it becomes a crime. If the company owner takes off with the money, the way I believe Von Mueller planned to do, it's a serious crime."

"So, we've got him for computer theft, with it on video, but don't know about her."

"Her husband goes to the hospital with head injury and I think she drove Von's sports-car to Bob's safe-house and shot him. Janet shares the same greed with Von as well as the tryst. It could be enough motivation for her to have joined in the war of the PhDs."

"Well, I'll put out an all points on his car I've got the license info I need, and we can arrest her on suspicion of those two incidents. At least we can get her off the street and start evaluation on her psychological condition."

"That's good," I said. "I'm going to find her husband and try to protect him till she's in custody."

"If you find him before I do call me, and I'll assign some officers to guard him."

We closed our cells and I restarted the car, turned onto Airport Boulevard and drove north to MLK and turned left on MLK driving west to Guadalupe and the University area. I found Wallace Norton at his home in Hyker Park. Knocking on the door first, I walked inside and heard loud snoring. He was leaning back in a blue wing-back chair with an ice-pack propped on top of his head. Using this chance to look around freely, I examined the bathroom and shower.

When I opened the shower door, I noticed the shower floor level was quite a bit lower than the bathroom. If even a short person had been standing on the bathroom floor as they struck Wallace with an object, it could have reached the top of his head. In the drawer of the cabinet I found a claw-hammer on its side toward the rear. I walked closer and saw a dried, red substance on the metal side of the hammer. I left it where it was and went to look at Norton's head. He was still snoring as I lifted up the ice-pack enough to see the top of the head and the wound could be seen through the thin bandage. It appeared to be consistent with the surface on the metal side of the hammer. I replaced the ice-pack and he still didn't wake up. There was a knock at the door and I answered it.

"You did get here first," Williams said. "Is he alright?" Motioning in the direction of the snoring he walked over to the wing-back chair for a better view of Norton.

I told Captain Williams what I had observed, and he went through the same motions himself and came to the same conclusions. We were discussing our findings, when my cell phone sounded off and I answered it. Steve Wingate had news from the hospital.

"If you could come back to Luling we need you," he said. "Bob's, OK, but I've got other problems."

"Bob's conscious and talking to you now?" I asked.

"Yes, he's talking plenty, but he didn't see who shot him."

"I'll come back to Luling," I said. "The APD has it fixed here."

"That's good," he said. "Bob was supposed to meet Nelda at the safe-house."

"Is she there now?"

"Yes, she's there now and will stay till you get there."

"Well, didn't you put Bob on the phone?" I asked. "He can talk to her."

"It's something I can't tell you on the phone, but you'll see when you get there."

Leaving the house, I was aware of Captain Williams taking the hammer with the blood evidence. Outside he talked to the officer he was going to leave with Norton. He said the man's name and we shook, but I was in a hurry to tie up some loose ends.

In about thirty minutes I was at the other end of 183 entering the Luling city limits. I drove through the center of the city, out the back way to the bridge and crossed over to the south shore of the river. Driving on the gravel road, now, I could see Nelda's red BMW parked in the front, in contrast to the primitive, rural setting. We still had plenty of historical type structures like this one, and Nelda's father was selling high-quality, non-petrol, sleek vehicles at reasonable prices. I parked in the grass next to her car, thinking it wouldn't be bad to go someplace in the sun like this on our vacation. A place where you'd climb log steps to the front door like this one.

Not having any idea about what was going on here, I reached for the doorknob and found it unlocked. Opening it, I went inside to find Bob Clauson stretched out on the long couch in the living-room. Nelda was seated on a straight-backed chair, holding a pan of water and wiping his head off with a wash rag. His head was propped up with several pillows and I thought he even looked worse than he did in the hospital.

"It's good to see you again, Dan," he said. "Thank you for all you've done, I'm OK, Nelda, get him a drink or something."

"There's a fresh pot of coffee," she said.

"That's good."

"Sit down there by the table, and I'll be right back with it."

She motioned for me to sit on an antique chair, like the one she had been sitting on. It was on the other side of a round, solid-cherry table, also antique by its vintage alone. I sat down and she brought in two cups, putting one on a coaster for me and the same with her own cup on the other side of the table by her chair. I told her mine was black without sugar. She also gave Bob a glass of ice-water and he took a big swig of it and placed the glass on the table. I thought he was doing well for someone who was just released from the hospital. As if he had read my mind, he seemed to pull himself together to speak.

"There's something I must tell you," he said. "I apologize for not telling it sooner."

He looked sad enough to start crying, but he didn't. I wanted to ask him what he was doing running around outside of the hospital.

"What, Bob is trying to tell you," Nelda said. "Is that the man in the hospital is Bob's twin brother, Don." That explained it. There were two Bob's, and one of them was a Don. Don was the one who got shot not Bob.

"Let me explain the rest," Bob said. "Don came into it, at the time I was being held captive. We had gone different ways since high school. While I had gone to college, he went to the theatre in New York on Broadway, and made a living of it as an actor. It was just a lucky break for me. He was between plays and decided to look me up in Crystalville.

"So, he found out you were missing, right?" I asked.

"Yes, and he went into the underworld in the greater Austin area," Bob said. "He played the part of a mob boss and enlisted the help of Bud Wilson and the others in a pool hall. We had never lied to each other and I believe every word he told me. He said that after they freed me from another vacant condo, Schmidt shot at the three men and chased them. They were still around somewhere hiding from him at that time. Can you imagine how Don felt? He had to leave there and find a place to hide me. He blames himself for bringing in those hoodlums to get me free though Steve was in on it too, Don had to get doctors and arrange everything to move me to the river house, at a more convenient location. I'm still trying to get over it, but at least I wasn't shot like poor Don was."

Telling him how I thought he had been living at the more northern location, at the river house on Pecan Street, made him go back and fill in the information at the spot where I was still confused. He told me that he and Don had no choice but to set up the deceptive act they had put on. Bob had been just in and out of consciousness, but was able to rest enough in the safe-house while Don and Steve followed his instructions and carried on at the job site. This made me understand how the molecular manufacturing had been proceeding at such a slow rate of speed.

Only Steve, Don and Bob were in on the duplicity. The two brothers looked so much alike that it had been easy, and, of course, Don was a good actor and his brother's best confidant.

"I didn't see the person that shot Don, either," Bob said. "I was asleep in the secret passage room and thought I heard something that woke me up. By the time I did get up and go into the rest of the house everybody was gone. Steve called me from the hospital and told me what had happened. I told him that Von Mueller was finally gone and what Nelda had told me about Harold Remkin's offer to us to return to Gladstone. That was when Steve confessed the whole story about how I was passing information back and forth between the two of them so they could do my job."

When I asked Bob if they had talked over the offer and was sure about going back to Gladstone, he said that they were. Don had said he was glad for them, and he planned to go back to New York as soon as he could get out of the hospital.

"Who else knows you've been directing things from this safe-house?" I asked.

"Only Steve Wingate," he said. "Don had to confide in him from the start, and Steve didn't see any other way to do it. I wasn't able to report for work at Comp-ease, but Don was, and we had a lot of three way cell phone conferences, to prepare Don for his big part of playing he was me."

"Good Steve broke down and told it all to James Lebeau today. Did he tell James how you were offered a job to work back at the Gladstone Lab with Nelda?"

"Yes, and he's been very understanding about it," Bob said. "By now he's running the entire show, and doesn't really need me, but Steve is going to stay." He picked up the glass of water again and drank the rest of it. Nelda refilled it and brought it back.

"I can understand, now," I said. "Why you didn't want to go back to Gladstone while Von Mueller was still there."

"I could see he wasn't trustworthy, when I worked there before." He sat up on the couch and drank some more water, seeming to be getting stronger.

"So when are you going to head back to Crystalville?" I asked.

"He was doing very well till Don was shot," Nelda said. "He's had a small relapse."

"I'm getting stronger now, and I'd like to go back today." he said. "It was physical weakness and getting tired and forgetting things that slowed us down." They just used that to help Don appear more subdued to play his part a little better. "When I came in the door, Bob was on the couch," Nelda said. "He got up to meet me and passed out."

"With all the excitement I didn't get around to fixing myself any breakfast," he said. "Nelda fixed me an omelet with toast and it stayed down."

It turned out that Bob was right about his own diagnosis. He'd been on the phone twice about his brother, and the first time Steve wasn't able to say that he had talked to Bob after the surgery, so Bob thought something was still wrong. He was getting ready to go over to the hospital, and was looking up the number of a taxi to call, when Steve called him back to say Don had come out of it. Bob insisted on talking to Don for a few minutes and so Steve took his cell to the bed and they did talk a short while. It seemed to exemplify the closeness of the two brothers. I had never seen them standing side by side and up close, but I had to assume they were identical twins.

When I thought about the relationship of twin brothers, it explained a lot of things about this case that I hadn't been able to figure out. I remembered Bob's recent self analysis, when he said his relapse had been a physical thing. Thinking of the uncertainty of Don's condition in the hospital had made Bob ill, and he had passed out. When Don had gone to visit his brother and found him missing, he had risked everything to find him and protect him, until Bob could get over the beating and the agoraphobia.

My thoughts about the duplicity of Bob Clauson were partially brought on by the fact that I had briefly talked to both brothers, separately. I didn't know there were two twins. The more personable version of the two was brother, Don, who was the actor. He had been such a good actor that he won me over and the duplicity went away. It seemed to me that we had better be careful of actors who become leaders in our government without the prerequisite experience in foreign affairs. Otherwise how would they know to accept or reject advice concerning correct foreign relations?

CHAPTER 50

In the real world things can work out if you survive long enough to overcome the obstacles. That was what I did, survived on this case until things started to break for me. I made some of the right decisions, but I didn't know what a crook John Von Mueller was till I got nearly to the end of this case. I was expecting to find some kind of terrorists lurking around, that I might not discover till the end of the case, but not a war of the PhDs. The latter was a very subtle kind of war that could be going on for a long time before you even became aware of it. I didn't think many PhDs would be like Von Mueller and be so insolent as to steal from his employer. It was good for me that he did, as it provided the kind of evidence I needed to make all the other charges stick.

Nelda and Bob had been planning to get married, and now that they were going to Crystalville Nelda was making specific plans and Claire and I were invited. What she had been holding out for had finally come about. The event was scheduled for next month and I thought Claire and I would attend. Under Nelda's special loving care Bob's condition had improved rapidly, and he claimed he was ready to travel back to his condo in Crystalville, after they made a stop on the way to visit Don at the hospital.

They drove together in Nelda's car and I met them at the hospital. We parked and went inside together. It was still visiting hours and Steve and James Lebeau were in Don's room when we went in. James had been standing next to the bed with Steve talking to Don. Looking up at us, James came to Bob with his hand extended, and smiled our way. I held back and let Nelda and Bob advance to make their introductions. I went over and shook hands with Don, feeling more at ease with him than I did with Bob.

"Thanks for coming over, Mr. Brandon," Don said.

"That's how I should have figured it out," I said. "Your brother, Bob, called me Dan, and when I talked to you, you always had a cold. The sniffles covered up the slight difference in your voices." Neither one had a wart on the nose. It had to be a smudge on the lens of the camera or some such thing. When I asked Don, he said that Bob had one taken off because of the fear of the big C.

We were both laughing when the others joined us. Everyone at least smiled, so I thought we must all be reconciled. We all had to agree that the Bob and Don show was the best brother act we had ever seen. Nelda thought that somehow the culmination of the last few days was bringing Bob out of the agoraphobia. I didn't think he had it, agreeing with his own diagnosis. It was a curable mental disorder and he could be back to normal soon if he did have any of it. He seemed highly motivated to get out of the house and go to the hospital to visit Don. He was confident about making the trip to Crystalville. They also planned to go back to the hospital, when Don was released, and take him too. The condo was where they were headed now and I was headed for Crystalville to check out with Harold Remkin, at Gladstone Electronics. It was the last time I was ever in Luling.

Telling everybody I'd talk to them later, I got in my car and drove to Austin on 183 and took the crossover to LH-1 as I always did. I got off at the Crystalville exit and took the private loop of Gladstone Drive to the parking lot. I parked and went inside at the front security entrance. Harold Remkin opened the gate by remote control and I heard his voice on the speaker saying for me to come on up.

He was waiting for me as I got off the elevator and we took the corridor all the way around to the left, where his office was still located.

"I'd return the key, but I had to leave it in the evidence room at the APD," I said. "They need it to compare it to the key they've taken from Von Mueller's house."

"Oh, we've got better ways of protecting our prototypes now," Remkin said. "There is a lock on the door, but Carl has keys to it in his office. His people take care of that now. I'm just keeping that key in the vault in my office to maintain the original system. Our legal consultants have advised me to do that, because of the impending case against John Von Mueller."

Presenting him with my itemized expenses and the total fee for my PI services, I sat down in the easy-chair next to his desk. He glanced over it for a few seconds and then went to the vault to get a company check book. When he made out the check he included a bonus amount to my total fee, which was the same figure as the amount he had added to Ray and Willie's checks.

"You don't have to add a bonus," I said. "You've already given it to Ray and Willie, as I asked you to do."

"I received a call from Bob Clauson," he said. "Bob accepted our offer, so you can think of it as a finder's fee we're paying you. And you'll be a witness and have to testify in the court case, so think about that too."

Not arguing any more about it, I put the check in my pocket, and gave a brief summary of how the case had ended. He said he had never heard Bob talk about having a brother and was surprised at that development as anyone else would be.

"Bob and Nelda are moving into his condo now," I said. "He was upset about the way his brother had been shot, while taking his place at Comp-ease. It seemed to be the jolt he needed to speed up his own recovery."

"He can get over it in his own time with us," he said. "He can run things from his condo as a consultant. That brother's secret stuff would not have been necessary at Comp-ease. If they had just told James Lebeau the truth about what had happened, I think he would have been just as understanding as I am about it. Of course he was when Steve Wingate told him in the end."

"We might think that, but Bob wasn't able to work with Von Mueller. He knew the man was a ruthless egomaniac and not trustworthy, so he felt stuck between places. He couldn't go back to Gladstone till he knew Von had been put away for a long time. Comp-ease was the only other computer company that was advanced enough for him to continue his nanotechnology, with the ultimate objective of setting up the first successful molecular manufacturing machine. Bob was so insecure that he felt he had to make it work there, because there was nowhere else to go at that time."

Wanting to be sure that Remkin would remember his own feelings of insecurity, brought on by the power struggle with John Von Mueller, I kept up that conversation till I thought he was getting it. It would help Bob in his new relations at Gladstone if Remkin could have good empathy with him.

"When I think about Bob working at Gladstone with Von Mueller, I can see how he would have been insecure anyplace," he said. "Good old Von had been courting some of the big stockholders, telling lies about me, and it almost worked."

"You're lucky he was self-destructive or he could have destroyed your career."

"If he wasn't, I might have destroyed my own career by killing him. I had bought a pocket revolver, and practiced with it on an indoor range."

"It's good you didn't do that, Harold," I said. "You'd have died in jail."

"Yes, but I wasn't thinking very well then."

"That's the trouble, at those times nobody thinks well."

Harold Remkin had probably been born an honest man in a long line of them. I could see that something bad had happened to him, when Carl told me he had changed and nobody could get along with him.

It took a while for me to learn that the bad thing that happened was John Von Mueller. I thought sociopaths were attracted to each other at times, and that was the case with Janet Norton and Von Mueller. He brain-washed her about the PhD war and she became his cat's paw. She attempted to commit the two murders of Bob Clauson and Wallace Norton because they were in her way.

Von had thought that if he eliminated Bob Clauson it would be the end of his competition in that business. He must have thought he had learned enough about the laptop to make it on his own.

Remkin tripped the mechanism at the front gate and was watching the picture on the screen in his office as I left. I got on the elevator and went down to the first floor. I was glad I didn't meet anybody else on my way out and have to make up any more of the stories about how I'd be talking to them later. After I walked through the gate downstairs it automatically closed and Remkin said "Take care now." I kept right on walking. Cops don't ever like to say goodbye. This case had me thinking I couldn't quit it but it seemed to end itself. Listening to my bosses on this case as it was closing, they had convinced me that there would continue to be more of these crimes related to molecular manufacturing. In the case of high-greed, as you might always have a chance to make some multiple copies of a brick of gold, I didn't think there would be a shortage of this sort of criminal.

THE END